THE FIRST,

THE FEW,

THE ONLY

THE **FIRST,** THE **FEW,** THE **ONLY**

How Women of Color Can Redefine Power in Corporate America

DEEPA PURUSHOTHAMAN

An Imprint of HarperCollins*Publishers*

 For information, address HarperCollins Publishers, 195 Broadway, New York, NY 10007.

HarperCollins books may be purchased for educational, business, or sales promotional use. For information, please email the Special Markets Department at SPsales@harpercollins.com.

FIRST EDITION

Designed by Bonni Leon-Berman

Library of Congress Cataloging-in-Publication Data
Names: Purushothaman, Deepa, author.
Title: The first, the few, the only : how women of color can redefine power in corporate America / Deepa Purushothaman.
Identifiers: LCCN 2021044347 (print) | LCCN 2021044348 (ebook) | ISBN 9780063084711 (hardcover) | ISBN 9780063084728 (ebook)
Subjects: LCSH: Discrimination in employment—United States. | Minority Women—Employment—United States. | Diversity in the workplace—United States. | Equality—United States.
Classification: LCC HD4903 .P87 2022 (print) | LCC HD4903 (ebook) | DDC 331.13/30973—dc23/eng/20211104
LC record available at https://lccn.loc.gov/2021044347
LC ebook record available at https://lccn.loc.gov/2021044348

22 23 24 25 26 LSC 10 9 8 7 6 5 4 3 2 1

For Ella, I hope you grow up in a world where you feel strongly you belong, as this is one of our most fundamental needs. I want you not to grow into the system around you, but instead to stand freely until you understand it and can use your power to determine where and how you want to root.

To the women of color who suffer in jobs and tasks outside the corporate system, your stories are important and your struggles often run deeper than the stories shared in this book.

To the women who are climbing in corporate now, power resides in each of you. Sometimes, you just need to be reminded it is there.

CONTENTS

RESEARCH AND STORIES IN *THE FIRST, THE FEW, THE ONLY*

All of the stories in this book speak to real-life issues and circumstances faced by women of color in corporate spaces. Some contributors shared their stories and wanted to use their real names. Some shared their stories but wanted to mask their identities. In these cases, I have created pseudonyms and changed details of their stories that might identify them to protect their identities. Finally, some of the stories in this book are a composite based on a number of contributors sharing similar narratives. For the composite stories, any resemblance to any real person with similar names and/or circumstances is purely coincidental.

INTRODUCTION

It was late September 2009, and I was sitting with my friend Walter from graduate school. We usually met in New York a few times a year to share our successes and our career hiccups. The bartender had just poured our champagne into two fluted glasses, and Walter raised his, saying, "Congratulations to both of us." In the last six months, we'd both made partner in our respective firms, him at his law firm, and me at a global professional services firm. It was an exciting time, full of possibility, and I was looking forward to celebrating and letting my shoulders down.

In many ways, Walter had been my biggest cheerleader. I often called him when I was stuck or wanted to celebrate big wins, and I always felt energized after our get-togethers. He seemed to always know what to say to perk me up and build my confidence. As we got to our second glasses and our meals came, he stopped me. I was far into a story, talking about the office politics I was navigating and the pressure I felt as a new partner, when he said, "Deepa, seriously, you have nothing to worry about. You are set. You are golden. You are going to move so fast in the partnership." I stared at him, fork in midair, confused. "What do you mean?" I asked. He laughed and took a sip of champagne. "You are a 'twofer.' *You* have nothing to worry about. I on the other hand, as a white man, am going to have to work hard to earn what comes next. You're going to skate ahead because you check so many boxes. Men like me, we're losing opportunities, but people like you, you can ride this wave."

I could feel joy draining from my body. At the time, I didn't even know how to articulate what I was feeling. I finished my meal in record time, and I never met Walter with the same enthusiasm or sense of safety again. In just a few words, he had identified my knot of insecurities around whether I was good enough and whether others around me questioned my worth and my skills. In just a few words, he had uncovered the confusion I felt about being a woman of color (WOC) at work.

That incident is one of many similar situations I encountered in my time within the corporate world. It was common to feel high and then low in the same moment. Something amazing would happen and then in the blink of an eye someone would say something ignorant. I'd feel the impulse to correct their words and their reality, while at the same time questioning my own.

I'm not alone in my experiences. Like me, many women of color I've met have scars from climbing the corporate ladder. We often question our reality. *Did that really just happen? Do they realize what they just said? Oh, they weren't expecting ME to show up!* On top of that, we don't have enough friends we can turn to because so many of us are the "first," part of the "few," or the "only": unique in the rooms we enter and in the places we stand as women of color at work.

"The first, the few, the only" is a phrase I use to describe women of color in the business world. Some of us are the first WOC in our families to go to college. We are sometimes the first to work outside the home or have a professional role. We are often one of the few women of color in our department or unit. And most of the time, we are the only woman of color in a senior leadership role in our company or organization.

As the first, few, and only, our path to Corporate America is almost always different from our white counterparts'. Early in my career, I remember looking around and seeing maybe one Indian woman ahead of me. I made up my mind that if I didn't see it, I would be it. In fact, I literally typed *you don't have to see it to be it* and kept it in an email. When I doubted my abilities, I would read that email to reassure myself. But finding that inner confidence is not always easy. We have to be creative to find ways to remind ourselves we belong.

There is only one path I see to surviving it all, dealing with situations like the one I had with Walter, and thriving. As women of color, we need to unearth our individual power. It is not power that comes from outside accolades that folks like Walter have been providing; it is power that comes from inside of us.

It is power collected from our culture, from our lived experiences, and from the traits that each of us possess as WOC. It is power we define for ourselves. And it is unique to each of us. Once we find our individual and innate power, we can feed it by creating community, and building collective power to sustain us at work and in life. This is how we survive the structure of Corporate America—which was not built for us or by us—and it's how we change the systems around us.

A LIFETIME OF NOT BELONGING

I spent most of my life in white places and spaces. The town I grew up in, Whitehouse Station, literally had "white" in its name. Most of the kids in my class—and, in fact, my whole school—were white. As a result, I spent my life questioning where I fit and where I belonged. My earliest memory of being different is of my mother, dressing me in Indian clothes, braiding my hair in "plaits," as she called them, and then sending me to school with a bindi on my forehead. I'm not sure I understood race and cultural differences back then. Instead, I internalized the pain and shame of being different, and tried even harder to fit in.

We never talked about race at home, yet it showed up everywhere, from which friends would invite us to their houses, to the racist remarks I heard at school, to dating boys who ultimately could not reconcile the color of my skin. It reached a breaking point late one night when I was living in London as a young graduate student. A drunk, white Englishman followed me from the tube to my flat late at night when there were few people around, making me fear for my safety as he called me the N-word and told me I was not welcome.

As the daughter of immigrants, I often felt even worse when we would visit India. I didn't fit in there, either, because I was seen as too Western. That confusion of not fitting in and adapting myself to navigate spaces followed me to work. All these experiences sank into my consciousness and stayed there, telling me: *You don't fit in. You don't belong.* As outsiders, especially if we don't grow up in communities of color, we spend most of our lives trying to fit in and feeling deep down we are different. I have spent my life living in competitive, high-performing, majority-white spaces, wondering why I was exhausted, confused, and drained of my power.

The details of our lives may differ, but if you are a woman of color, you have probably had to deal with these same issues around belonging—whether it's because of your race, ethnicity, class, gender, sexual orientation, disability, or any combination of these factors. We don't see ourselves in our teachers and advisors, or on television and in the movies. Many of us have been taught that being successful means toning down our looks, our dress, and our personalities in order to fit in and not stand out.

Growing up, we've watched the "system" take power away from us as girls of color and give it to others—like little white boys. They are given messages that they will succeed, be important, lead. We often don't hear or see those same messages. Leadership is not made in our image. And, consciously or unconsciously, that can limit what we believe we can become and how we see ourselves in the world.

Power, as it has been defined to date, has never included WOC. We are told that leadership, success, and power are static, established, and universal. If we do end up leading, most of us follow models that feel incongruent to who we are as WOC, ultimately chasing ideals that will never work for us. Eventually we rise to the top and find ourselves performing, maybe even excelling, but not feeling powerful.

Because most of us were taught to value success and achieve the so-called American dream, we often endure microaggressions—and sometimes worse. We've worked so hard to get on and move up the ladder that we don't have a chance to stop and really study corporate culture, to question and change it. We have been taught to be grateful and thankful for being included, so when issues arise, we often don't even consider pushing back. We haven't been given the language or the tools to work against the status quo, and we have no blueprint for restructuring the system.

If we do find the courage to push back, the system stops us in our tracks and will try to uphold and spread a definition of power that works for it rather than us. Any definitions of power that are different from the white male heteronormative standard tend to be dispelled.

Behind Corporate America's veneer of supporting inclusion, it has never fostered true equity, especially for women of color. As part of the first, few, or only, I want to give us permission to question everything, and to redefine power in a way that suits us. This means figuring out for ourselves what makes us feel power-*full*, rather than using the playbook most of us have been handed—one that has left us feeling depleted.

MY WORK AS A WOMAN OF COLOR

On paper, I am what you might call an ultimate insider. My schooling and the networks I've developed have allowed me to sit at the table with some of the most prestigious and influential people on the planet. I have lived in the belly of the beast, spending over twenty-one years inside a corporate structure, where I served some of the most iconic global technology and telecommunications clients. I became a consulting partner at Deloitte, the largest professional services firm in the world, and was one of the youngest and the first Indian woman to do so. In addition to advising clients on how to transform their businesses, I led Deloitte's renowned Women's Initiative and was responsible for inclusion strategy for more than one hundred thousand employees in the United States. It was one of the highlights of my career.

So when I decided to leave midcareer during that upward trajectory, those around me were surprised. When I said I was shifting gears to focus on women of color issues, they were really confused. And when I told my trusted circle that I wanted to write a book for and about women of color, many of them asked: "Why?" Why would I put aside more than twenty years of business experience only to be defined by something so contentious? Why would I want to dedicate so much time to such a taboo topic? My inner circle explained that it wasn't lucrative, and it wasn't easy. People told me: "Race is such a hard topic. You won't get any credit if you get it right, and a mess of people will be waiting on the sidelines to tell you everything you did and said wrong."

But I knew I wanted to change the narrative of older white men representing the echelons of power in America. I wanted to show that leadership could look different, and in order to do so, more of us needed to tell our stories. Over the last few years, even under the cloud of COVID-19, I did just that.

I left my job, started research on structural racism with the Women and Public Policy Program at the Harvard Kennedy School, and shifted my entire focus to the issues WOC face in corporate spaces. In 2020, I started my company, nFormation, with my business partner and former executive coach, Rha Goddess. It's a platform that creates a safe, brave, and new space to help professional WOC find their power and come together. We provide leadership programming, coaching, and innovative placement services for WOC. We created it for the first, the few, and the only, who are often isolated in their positions. Our goal is to help WOC navigate challenging situations, develop their individual and collective power, and create community across WOC so that we can change how the table is formed and redefine leadership.

According to the *Harvard Business Review*, by 2060, the majority of American women will be women of color—and, in turn, WOC will most likely comprise most of the workforce. Already, we generate $1 trillion as consumers and $361 billion in revenue as entrepreneurs, launching companies at four times the rate of all woman-owned businesses. We are growing in size and power, and it feels like our moment. I see power in us coming together. We are a growing force.

Let me clarify this before we go further: I clearly identify as a woman of color. I also acknowledge identity is fraught with confusion, politics, and history, especially in the United States. In this country, power has always been tied to who and what you are.

WOC are not a monolith, and we each bring our different cultures, identities, and experiences to the fray. I know some people don't like the term "women of color" because they think it dilutes the identity of being, for example, Indian, Black, or Indigenous. I recognize some women's experiences are different and distinct. So, where I see differences by race, I note them. You will see places where I acknowledge when something is, for instance, unique to Black women. I will mention ethnicities such as Japanese or Chinese and Colombian or Costa Rican where I can, but use race or broader categories like Asian and Latinx where I can't.

But I am comfortable using this term because we've all been burdened by shared experiences in rising in the workplace. I see the patterns we face in navigating structures like Corporate America. We often feel like we must leave some aspect of our identity behind as we badge through our turnstiles and head to our desks. And as Rha, my business partner—and also founder and CEO of Move The Crowd, entrepreneurial soul coach, author, and self-proclaimed corporate refugee—says, "We have been conditioned to button up and armor up to survive."

Professor Efrén O. Pérez at UCLA suggests that we should use terms like "people of color" (POC) not to take the place of our whole identity, but as a super identity, acting as one more way to describe our experience—a term that speaks to not being white in structures that revere whiteness. For example, I can call myself American and still see myself as Indian. Pérez suggests we can keep our racial or ethnic identity and see terms like POC and WOC as additions, when needed. His research also shows that using terms like POC (and, I believe, WOC) builds momentum and helps us rally behind shared goals of changing the world around us.

I use the term "women of color" in solidarity. There is power in coming together when talking about evolving workplaces. We need to stand together as one to create a movement for us all.

THE STORIES WE NEED TO TELL

I wrote *The First, the Few, the Only* to help all of us who are working in Corporate America—though the powerful stories may resonate well beyond corporate structures to anyone who wants to help WOC by co-conspiring and reimagining workspaces to have greater access and parity. I use the term "co-conspirators" throughout this book instead of the term "allies" because it's more action-oriented. We need allies to do more than cheer us on from the sidelines. Activist and author Alicia Garza, in an interview titled "Ally or co-conspirator?," says, "Co-conspiracy is about what we do in action, not just in language."

It was also written for leaders who want to gain insights to challenges WOC navigate. While many companies and white leaders have a lot to fix and maybe even atone for, they must do that work on their own. In the meantime, this book is a resource for WOC who are trying to survive, thrive, and find their power. It was written for women who are entering the workplace and want to understand the dynamics to come; women who may be questioning their path or calling, wondering why they feel isolated, tired, and confused; and women sorting through the challenges they face after breaking down those barriers.

Women of color get pushed and pulled at each step on the corporate ladder. Entry-level WOC are trying to learn how to fit and adapt to their job and company, and they do not always have the power to make their voices heard. Midlevel WOC are caught wanting to grow and make a difference, but they can become jaded by the lack of change they're able to inspire. And senior-level WOC are often the loneliest and the most embedded in their companies, with the most pressure to conform.

You will read stories from women at each of these levels. I met and spoke to women in cities across the country, across industries, of all ages and levels, and from a variety of races, ethnicities, and countries of origin so I could understand, firsthand, what it felt like for them to be women of color in their companies and workplaces. I met some women for one-on-one meals, some women in groups, some on the phone, and others over platforms like Zoom. Some of these were women I knew well, and others were women I had admired from afar. Each had unique experiences and yet also common threads to share.

As they shared their stories, I began to see that we are in a rare moment of opportunity for women of color. The women I met are intelligent, driven, and even rebellious in their thinking. They are fighting hard to get and stay at the existing table and are prepared to talk about how to flip or destroy that table to create a new one that will suit all of us. They want to alter existing narratives, change the game, and lead in a brand-new landscape.

Having now interviewed more than five hundred women of color, and spoken to a number of academics, change agents, human resources leaders, and co-conspirators from white leaders to men of color, I can safely say that whatever your story is, you are not alone. Throughout *The First, the Few, the Only*, I weave in some of my own history, but I also tell a variety of other women's journies so you can see how universally similar our paths can be. This isn't a book that's meant to give you all the answers; its aim is to help you find your own narrative and lead you to ask new questions about the spaces around you. You'll find stories in these pages that will speak to you and may even have you screaming at the page. Most important, I want you to feel hope and empowerment, and that change is possible.

The book is organized into three sections. Together, they create an arc that starts with individual power and flows into collective power—because you cannot have the latter without the former.

The first section, **Find Your Power**, explains how to reimagine traditional belief systems and see through delusions that strip you of your power. You will discover the gap between what corporations *say* they stand for and what actually happens inside. Most important, you will learn to shed unhelpful messages you have been taught so that you can carry your wisdom forward and come into your own personal power.

The second section, **Feel Your Power**, strengthens your intuition and innate traits so that you can define your leadership style in ways that are more authentic to you as a WOC. You will learn how to listen to your body when it tells you to pause and rest, when to push back on the extra tasks we are asked to perform as WOC, and how to conserve your energy when facing microaggressions and fighting structures around you.

The final section, **Forge Our Power**, examines the construct of power itself. You will discover what is possible when we band together, and how our collective experience can transform Corporate America. You will learn when to exit your workplace and when it's time to explore new opportunities. And you will find out how to play the game and change the game if you stay where you are.

When I first started writing this book in early 2020, I thought I would need to explain how race was a real factor in corporate spaces and why race-related conversations were relevant within company walls. But then 2020 happened, with COVID-19 and the resurgence of the Black Lives Matter movement, and the world and work changed. In less than twelve months, the entire landscape shifted under our feet, and conversations on race at work have evolved. We are on the cliff's edge of radically new ways of working and doing business, and I believe real change may finally be possible. And we, as WOC, are uniquely equipped to remake and reimagine what comes next.

Years ago, someone told me about a Zulu word, "sawubona," which means "I see you; you are important to me and I value you." The women of color I met said they wanted this for themselves—to be seen and valued for who we are rather than how we are interpreted by others. This is what I want for you. I want us to see ourselves clearly in our own eyes, with our own definitions, so we can step into our leadership roles and stand together in our full power.

PART I

FIND YOUR POWER

CHAPTER 1

THE DELUSIONS THAT HOLD US BACK

In order to find our power, we must first break through the delusions we are forced to ignore as women of color operating in corporate spaces. Delusions are so-called rules that have been set up around us. They are stories we have been told to make us believe and act a certain way. These delusions are the unwritten rules of how Corporate America works. We need to see them and recognize them for what they are: antiquated rules that have been used to keep corporate structures static, and to keep power in the same hands, for centuries.

In October 2020, I called my friend Vernā Myers, the VP of inclusion strategy at Netflix, to ask her advice about my new company, nFormation. Vernā had been outside of Corporate America for twenty-two years, making a name for herself as an inclusion consultant. Now she was back in corporate, navigating through competing demands but loving her role. I joked with her that at the same time I was leaving Corporate America, she had returned.

We started talking about why so many women of color struggle at their companies, and she said, "Deepa, Corporate America wasn't designed by women of color." I was expecting her to talk about challenges with sponsors or issues with not having enough senior role models. Instead, she talked about airplane design.

As late as the 1970s, only 2 to 3 percent of Boeing engineers were women. Vernā believes that if women had been involved in designing airplanes, the interiors would have been drastically different. As women, we would have somehow figured out how to design a stowage that didn't require us to stand on our tippy-toes and lift a heavy bag over our heads into that small compartment.

"I'm pretty tall," she said, "but most women are shorter than me. And for many of us, upper-body strength is not our strong suit." She then added, "Who wants to feel unwelcome, not strong enough, like they don't belong, within minutes of entering a space?"

As a petite, five-foot-two woman, this struck a chord with me. When dealing with my luggage, I often wish I were taller or wonder why I packed so much. I always feel a wave of relief when my bag is tucked away, and I can finally sit down, knowing that stress is behind me. Many of us feel that same sense of dread. The stowage wasn't designed for us, yet we start to believe *we* are the problem. Why do I feel like it is my deficit when I can't put my luggage up high?

Vernā says the same thing goes for the workplace. "We weren't at the table designing the original workplaces," she said. "As a result, a lot is missing or difficult to succeed at without a whole lot of effort. Corporate America was structured in a way that advantages some groups and disadvantages others because it was built with men at the center, men as the norm, and white, cis, straight men at that."

The worry that something is wrong with *us*, rather than the design, the system, or the process, is a delusion. In reality, certain principles were missed because we weren't part of the team. Workplaces need to be redesigned and reimagined completely for POC and WOC.

There is overall denial that we are working inside a system of delusions. We have been taught and coached to just push through them. We've been told to just work harder and be better, and we will be rewarded. When we raise our voices or question delusions that make our skin flush with anger or make our stomachs lurch, we are told we are being overly sensitive, or seeing challenges that "aren't there."

The truth is, Corporate America has never actually fostered true equity—especially for women of color—and company cultures aren't set up to support us. In February 2021, *Harvard Business Review* ran an article titled "Stop Telling Women They Have Imposter Syndrome," and authors Ruchika Tulshyan and Jodi-Ann Burey noted, "We don't belong because we were never supposed to belong. Our presence in most of these spaces is a result of decades of grassroots activism and begrudgingly developed legislation. Academic institutions and corporations are still mired in the cultural inertia of the good ol' boys' clubs and white supremacy. Biased practices across institutions routinely stymie the ability of individuals from underrepresented groups to truly thrive." It's not our deficit that makes us doubt our abilities and drains our confidence. The corporate structure that was created by patriarchy won't allow us to feel powerful because, for the structure itself to be in power, it must diminish ours.

We need to put words to what is actually happening and why many of us feel lost and battered as women of color in Corporate America. We need to see these delusions and talk about them if we have any chance of changing them. The ten delusions below are at the heart of Corporate America, and they are blocking most women of color from feeling truly included and stepping into our power at work.

DELUSION 1: "WE CAN'T FIND YOU."

The idea that the pipeline is broken and "we just can't find you" is a myth. Data shows people tend to correlate and congregate with others who are just like them. In 2014, the Public Religion Research Institute's Robert P. Jones noted that overall, white people's social networks were actually 91 percent white, and 75 percent of this same group had entirely white social networks without any presence from people of color. This suggests that white recruiters and HR leaders will also have significantly white networks, which therefore cultivates a white pipeline. Many of the individual WOC I met could easily name more than one hundred qualified people in their networks. White leaders simply are not looking in the right places. They are also creating leaks in the pipeline inside their own organizations.

Lyn, a fast-tracked, Chinese American female partner at a top firm, used to work with Janet, a Singaporean woman she admired. Janet was a top-rated senior, a level below partner, and was brilliant at what she did. But Lyn's mentor, David, a senior white male partner, confided over drinks that Janet would never get ahead because she spoke heavily accented English and just didn't fit in with the other partners at the firm.

When she heard this feedback, Lyn went silent. Initially she felt great shame and didn't say anything to Janet, but eventually Lyn encouraged her to transfer to the firm's Tokyo office, where Janet made partner in just two years—leading Lyn to question how promotions were made in her firm. Why had Janet been able to make partner overseas but not in the US? Was it because Janet was an immigrant WOC and the white male team making promotion decisions dismissed her ability to lead in her own way? This is how leaks in the pipeline come about.

Consciously or unconsciously, David was gravitating to what was familiar. Lyn's and even Janet's experiences are not unique. When leaders feel comfortable only around people they identify with, where it feels "easy," we end up with fewer WOC rising. Leaders like David might say, "We simply cannot find qualified women of color candidates, especially for leadership positions, and we don't have enough qualified internal talent." This is not a pipeline problem. It is a finding and keeping problem.

When McKinsey and LeanIn.Org published their study "Women in the Workplace 2020," they showed white women and women of color combined comprised nearly 50 percent of the entry-level employees. Both groups shrank as they approached the C-suite, but six times as many white women ended up in that C-suite as women of color. Companies are not looking in the right places, and most are not addressing all the ways in which they force WOC out.

DELUSION 2: "JUST BE YOURSELF."

Many of us begin our careers with an optimistic view of our power. We believe we will be able to stay true to ourselves while being successful and making change. One of the biggest delusions is that Corporate America doesn't require us to conform and assimilate to be successful. This myth has allowed companies and leaders to maintain the status quo while asking us to fit in. Arguably, the penalty for not adhering to white norms starts well before employees are hired.

The well-known résumé study by Marianne Bertrand and Sendhil Mullainathan in 2003 set out to determine if experience or race mattered more in the workplace. They sent résumés in response to help-wanted ads in two major American cities. They found that a white name yielded as many callbacks as an additional eight years' experience from a person of color. In a 2016 study, "Whitened Résumés: Race and Self-Presentation in the Labor Market," Kang, DeCelles, Tilcsik, and Jun found that "whitening" résumés by changing names and erasing certain memberships or experiences that revealed racial or ethnic affiliation resulted in more callbacks from employers. Even when companies claimed they were looking for diversity, they showed preferences for the whitened résumés, suggesting that applicants of color were correct to assume racist hiring practices.

The challenges get worse once we walk through the corporate doors. Companies should actively be asking women of color what's causing us to stand still, shrink, or opt out. And the answer is statements like the following:

"You need to dress more professionally. Don't wear your hair that way."

"You are being too loud. Don't be so emotional and excitable."

"We don't see you at the next step. You need more executive presence."

"We don't really know who you are. You need to fit in better and get along with others."

Every day, we try hard not to let our differences and varied experiences take up space and change the way we are perceived in the environment around us. We try to normalize our experiences of conformity. As we will talk about in Chapter 5, many of us code-switch and adapt who we are to fit in. This causes a lot of confusion, and this gray matter is part of what it means to be a WOC at work. The silent pressure to conform is a form of structural racism. And it happens at all levels of our careers. One of the less-tenured women I spoke with, who had just entered the workforce, doesn't always feel like she has the choice of nonconformity if she wants a paycheck. She's still sorting through basic ground rules and expectations of the workplace, even wondering if just wearing red lipstick will "negatively affect others' perception" of her professionalism.

In order to avoid causing "discomfort" to our coworkers, we try to tamp down what hurts us. The day after George Floyd's death, many Black employees felt they had to start yet another workday "as usual," and many Asian women weren't sure if they could bring up the anti-Asian bigotry and violence that occurred in 2020 and 2021. As Ella L. J. Edmondson Bell and Stella M. Nkomo wrote in their introduction to *Race, Work, and Leadership: New Perspectives on the Black Experience*, "Employees do not leave their race or racial beliefs at the entrance when they enter the workplace." Yet we are expected to bottle up emotions, get the work done, and pretend that the constant news doesn't frighten and anger us.

The deeper paradox is that the more we try to fit in, the more disempowered we become, and the more disempowered we become, the less we can feel true belonging. By working to fit into existing power structures and establishments, we lose a lot of what makes us who we are. Many of us feel empty, hollow, and diminished.

The message is that we are different from the white male leader who created the corporate ideal of leadership, and the delusion we are told is that we need to be more like him, when in actuality, our differences are part of our power.

DELUSION 3: "JUST WAIT."

What is interesting about Delusion 2 is that most WOC I met felt heightened expectations to conform the more senior they became. When women got past the senior director level or the equivalent, they felt even more pressure to fall in line, to not stand out as different from their peers.

When I was in New York, I met Maci, a newly retired CFO at a public company. As a WOC, she had reached the level of status and seniority that many of us aspire to. She had always believed success meant getting to the table. On her way there, she thought she could be her own person, use her voice, and finally fight the system from the top. If she "just waited," she could eventually "do it her way." But once there, in order to be accepted and included by her male C-suite colleagues, she was expected to play by a *new* set of rules. The challenges that arise at more senior levels may also have to do with the concept of "pet to threat," which a number of women spoke about. "Moving from Pet to Threat: Narratives of Professional Black Women," a 2013 study by Kecia Thomas at the University of Georgia, suggests Black women are embraced early in their careers, but as they gain skills, confidence, and ambition, they begin to be viewed as threats.

Maci often felt she was barely tolerated in the executive ranks as a female CFO, and that when she was accepted by her peers, it was on the condition of being silent and fitting in. When racist comments were made around her, she learned quickly it was better for her career and her stress level to just go with the flow. "In truth," Maci told me, "I wasn't always forthright about who and what I was as a light-skinned Black woman. I often stayed quiet on topics of race."

Most of the women I interviewed believed erroneously that, once they were a leader in their organizations, they could be more of themselves. But once they reached the top, they were more invested in the company, and it was even harder to speak up, stand out, and be different. One woman I interviewed shared, "They promote you, but ultimately, what they want is a skirt-wearing dark-skinned white leader with an accent."

We have to let go of the fairy tale of "one day" and realize that no matter what stage we are at in our careers, each action to conform is actually a decision we make, even if it might not feel like it in the moment.

DELUSION 4: "I DON'T SEE COLOR."

Many white leaders I met in 2020 said that they take a "color-blind" approach to hiring and evaluating employees. One actually said, "I hope we won't need things like inclusion one day because we will move beyond race." This perspective misses the point that race and ethnicity have in fact been built into our structures and our systems.

America has built its economy on laissez-faire principles and the false idea of meritocracy. These claims of meritocracy that defined American business culture have perpetuated the issues women of color face, like unequal pay, bias in the review process, and lack of senior WOC in leadership roles.

When we falter or don't rise, the delusion of meritocracy puts the blame on WOC. But the idea that Corporate America is a meritocracy is a setup. Michael Sandel, a political philosopher at Harvard Law School, says that the more we view ourselves as self-made and self-sufficient, the less likely we are to care for the fate of those less fortunate than ourselves. He says this view of personal responsibility makes it hard for us to imagine ourselves in other people's shoes. Therefore: "If my success is my own doing, then failure must be their fault."

For decades, we've been coached to believe that the meritocracy principles that underpin Corporate America don't allow for racism. We were taught to believe that Corporate America and capitalism are color-blind. Most of our white colleagues didn't fully accept that the experiences of people of color were fundamentally different from their own. Until the summer of 2020, most employees I spoke with said it was unthinkable to even call the workplace racist or talk about systemic racism openly.

In the summer of 2020, Kareem Abdul-Jabbar wrote an op-ed for the *Los Angeles Times* where he explained that racism exists everywhere, "like dust in the air." We are so used to the "dust" that we don't notice it until the sun shines through and there it is, hanging heavily. There is growing acceptance that bias and racism exist in our operating structures and even governing structures, and that dust has settled in Corporate America, too.

As WOC, we need to know there are limits to the idea of meritocracy because we face them every day. The truth is, your hard work alone will not get you through. The meritocracy ideal doesn't account for the airplane analogy. Yes, I am on the plane, but my experience of getting my luggage overhead is hugely different from that of the five-foot-ten-inch man sitting next to me.

DELUSION 5: "PLEASE SHARE YOUR THOUGHTS."

During the summer of 2020, many companies launched forums asking employees of color to talk about their racialized experiences in America and within the workplace. The women I spoke with had varied opinions on the toll this effort took. Some felt it placed an extra burden on people of color to share their points of view. It put us in the hot seat yet again, pointing out our differences. Other women shared that it was hard to be candid about grievances when many companies were making head-count cuts while asking for feedback. Some of them feared retaliation for their candor. These discussions also perpetuated the pattern that our voice and expertise are often solicited for the "greater good," but we are not always fairly compensated.

These soundings run into a larger delusion that pervades Corporate America. Managers and executives will say things like "I really want to hear your opinion. Your voice matters. I want to know what you have to say." But most of them don't really want you to say anything unless it's positive. If what you say causes a ripple, that makes *you* the problem, not the company's outdated ways of working. The truth is, the corporate system can reward groupthink and inhibit the forward progress that most companies say they are encouraging.

DELUSION 6: "THAT'S TOO POLITICAL."

In June 2020, about a week after the murder of George Floyd, I got a call from a white, male CEO who had been given my name from a friend as an "inclusion expert." His communications leader had crafted a note to go out to his executive team about race in America. He was worried the tone wasn't right. That it might be too controversial. He asked me if I would look at it and give my opinion. Since it wasn't a big ask, and I wasn't taking on formal inclusion work at the time, I said I would look at it without a fee.

I read the note. It was benign, if not on the verge of boring. There was nothing special in it. It didn't reflect any of his own views and it didn't mention race. When I pushed him to share more, he finally said, "These ideas are too political. I am so afraid of saying the wrong thing." I convinced him to be braver, but even after the changes, the note was lacking any real insight or true leadership.

What I experienced with that CEO is the same position many leaders take: not wanting to be controversial. There is a belief that we can fix inclusion in Corporate America without talking about the hard stuff and making people uncomfortable. In *Race, Work, and Leadership*, Bell and Nkomo point out that "the explicit discussion of race and organizational leadership is still considered taboo or irrelevant in many business circles."

Studies like the American Community Survey show that, despite increased diversity at the national level, racial homogeneity still exists at the neighborhood level, with the average white person continuing to live in a neighborhood that is 71 percent white. That means one of the biggest opportunities to shift culture is by talking about race at work.

The truth is, we need to alter and shock company cultures in new and uncomfortable ways if we want our workplaces to reflect the change we are trying to make in the world.

DELUSION 7: "DIVERSITY, EQUITY, AND INCLUSION WILL FIX IT ALL."

In the fall of 2018, I was at a dinner with a room full of women of color. We were talking about why inclusion felt challenging to so many and why, even with so much focus over the last few years, results still fell short. The women unanimously agreed that DEI efforts don't address—and cannot single-handedly solve—the real, underlying race-related issues in the workplace. These efforts do not hit hard enough at a toxic workplace culture that makes women of color want to flee.

Dara, an organizational design specialist based in Boston, is the only Black woman on the partner track in her firm. She developed the firm's new plans around inclusion as a development measure. It was a way to showcase her leadership skills and to present at some of the partners-only meetings. In her decade at her firm, this project was one of the things she was most proud of. But when she asked the firm to let her implement these plans and lead the effort, her sponsors tried to talk her out of it, saying, "You can dabble in inclusion, but don't specialize in it—that would limit your career." They added, "Make it your minor, not your major. Inclusion isn't part of our core business focus." Dara shared that it deeply worried her that inclusion wasn't the high-priority topic the firm was proclaiming it to be, both internally and externally.

According to LinkedIn data, the number of people globally with the "head of diversity" title more than doubled (107 percent growth) over the last five years. A few women of color said it was yet another example of expecting WOC, and especially Black women—who are the ones assuming many of these roles—to fix broken systems with little support or resources. Others, who would not go on the record, shared that they think chief inclusion officers are a bad idea because it seems to become that single person's responsibility to fix inclusion, when it should be a C-suite issue instead.

These days inclusion is stated as a company priority and a CEO and board agenda item, but unfortunately, it's still just lip service within many companies. One of the white male leaders I interviewed went so far as to call company inclusion programs a form of "propaganda." It made his company look good, but it wasn't core to their business.

DELUSION 8: "KILL OR BE KILLED."

In Corporate America, the definition of "power" can often mean being aggressive, maybe even ruthless, and having power *over* others in corporate settings. Because of this, most women of color I met have a negative association with the idea of power, saying it is corrupt, unfair, and intrinsically uncomfortable. The idea that power must be seen this way is a delusion, and the idea that this type of power should be rewarded is an even bigger delusion.

In order to rise and thrive as women of color, we need power and leadership to evolve for the betterment of us all. So many women I met are leaving the corporate world to start their own businesses because many don't see a way to have power *and* be themselves within Corporate America, to enjoy their work *and* to realize their ambition. We need to change this. One twenty-five-year-old I met, who was still in the first two years of her corporate training role, shared, "I don't want to be what I see."

If we want more diverse leaders to aspire to corporate spaces, leadership must be redefined to include equality, empathy, fairness, openness, and heart. Power isn't about becoming a CEO—it's about being able to be true to who you are, what you believe, and what you stand for.

DELUSION 9: "CAPITALISM TRUMPS ALL."

In the wake of COVID-19, the growing dialogue on equity, access, and sustainability has opened the door for a discussion on the delusions that hold up capitalism. As more people enter the workforce, including women, immigrants, and future generations of workers, companies have simply pulled them in and allowed them to become overworked, underpaid, and burned out. I spoke with Rebecca Henderson, Harvard Business School professor and the author of *Reimagining Capitalism in a World on Fire*. She shared, "We believe one model [of capitalism] will work for all people." According to Henderson, we assume that everyone is competing equally, and that "externalities like the environment or the impact on humans don't count"—but many of us don't realize that these are actually false ideas. Other scholars I spoke with suggest capitalism in the United States is inherently flawed because it is based on a history of enslavement. It was built on the exploitation of people, and that cannot be ignored or erased.

I have a wise teacher, Daniel, who has studied ancient philosophies at length. He said it struck him as absurd that there is no dollar figure attributable to a living tree. There is no associated cost for cutting down a tree, yet there is a cost for a two-by-four at the store. He asked, "How is that possible? Is the currency we are using simply wrong?" Does capitalism supersede everything else?

There is a delusion at work that shareholder value drives all. Overworking and outperforming are our North Stars. We have been living as though we can extract value from the environment without recompense. All these delusions are being called to the center of mainstream discussion. The singular focus on growth works for companies, but does it work for society and for us as individuals? Does it work for women of color? The truth is, we can't talk about capitalism without talking about power. And challenges in power have been created and exacerbated by capitalism.

DELUSION 10: "YOU GOT WHITE-MANNED."

In the fall of 2018, the Aspen Institute hosted a private dinner in New York for inclusion leaders. That night, I heard a term I'd never heard before, and it perfectly describes the way many white men are experiencing the world right now. A male inclusion leader explained the term to me like this: "When Harry gets the job Jim wanted, Jim may be disappointed, but he doesn't blame the system. If Jamila gets the job Jim wants, there's a special word for that: 'You got white-manned.'" The term describes the belief that advancing women of color will take seats away from others, especially white men. This idea of scarcity is the most important mind shift we all need to make if we truly want to change traditional definitions of power.

When you are accustomed to privilege, equality can feel like oppression. And, yes, some white men may feel the pinch of their privilege shrinking. Some even feel inclusion programs put them at a disadvantage. In 2017, Ernst & Young surveyed one thousand workers and found that more than one-third of respondents felt their companies' focus on diversity lessened the focus on white men.

Remember my friend Walter, who thought my career path was set because I'm a woman of color? Businesspeople like him often see inclusion as a zero-sum game: if you get a promotion or an opportunity, that means someone else doesn't. Isabel Wilkerson addresses this in her book *Caste: The Origins of Our Discontents* by describing a deep-seated consciousness at play—a belief that as people of color rise in the ranks, the white predominant majority will lose.

We have all been living with the delusion that if we don't outshine, outwork, and outdo the next person, they will get the opportunity and the seat. I think this is the most important delusion to eradicate if we are going to create better work cultures and have greater equity. We can't have others, especially white men, believing they will lose something as we rise. The pie doesn't just have to be redistributed—it can always grow.

MOVING PAST THE DELUSIONS

Working in a world full of corporate delusions can push many of us to the point of sickness and exhaustion. These spaces weren't designed for us, don't see us, and don't make room for us to rise. Some of us contort ourselves to these delusions in order to perform and get ahead. Others must opt out by finding new opportunities or creating our own businesses with, hopefully, healthier workplace cultures.

No matter what, it's time we take these delusions down. These antiquated and universal myths are holding up Corporate America, but the system will *not* crumble if we remove them. In fact, these delusions are holding corporations *back* from hiring and promoting new talent, inspiring enhanced innovation, and even helping our planet to survive.

When you are ready, you can pull them apart and create a new way of looking at the world using your own voice and the truths you determine for yourself. How do we finally break through the delusions, take back our personal power, and move forward and effect change?

As the next two chapters explain, we must first take stock of our upbringing and cultural beliefs, deciding what to leave behind and what to carry with us. Then we need to focus on the strength and the power we have as women of color to change and rebuild our existing structures. It's the only way we can redesign those airplanes so that they'll work for everyone.

CHAPTER 2

SHEDDING MESSAGES THAT HARM US

Personal delusions are just as dangerous as the corporate delusions we discussed. They keep us bound in old ways of working and living. As women of color they make us feel invisible and they don't allow us to be our full selves. We rarely question them, and even if we do, we are told that that's just the way things are "always" done. We need to find the strength to break through them so they can no longer stop us in our tracks and keep us in our place. We can decide to leave them behind and rewrite our own narratives, creating room for ourselves to be in our full power.

When I first met Lisa Sun, she commanded the room and welcomed the attention that came with it. While she could be loud and boisterous, she also presented as solid, grounded, and wise in who she was and how she wanted to be seen. So when she told me that liking herself and finding her voice were relatively new to her,

I was floored. She seemed like the definition of unapologetic. I assumed she'd always had a high level of confidence and was born with bravado. What I learned was that, to get to where she is, she had to first shed the unhelpful messages in her life.

As a Taiwanese American, Lisa was taught to be loyal, obedient, and high-achieving. Her parents raised her to quiet her passions and instead focus on school and high-paying jobs like a good immigrant girl. Eventually her own metric became unattainable—she wasn't happy even when she received an A++ in AP Chemistry.

Growing up, Lisa went through the assimilation most Asian women go through, and this continued into her work, where she still believed she needed to act and dress like the people around her. She hid parts of herself and played their game. When people asked her to get coffee, treating her like a brand-new staff member instead of realizing she was a tenured professional, she would feel ashamed. She internalized these comments and worked harder to conform and fit in.

Late in her corporate career, it hit her that she would never be an accepted member of the white male club. After much reflection, she understood that to actually break into the club, she had to share more of who she was. She realized that shame and vulnerability were opposite sides of the same coin. She would never become a white man, but if she let herself be open and vulnerable, her colleagues would feel an actual connection to her, and they might accept her more. That's when it clicked for her: the more she shared, the more she would shine. She needed to unapologetically be who she was and leave her shame behind.

"Give people a window into your story," Lisa suggests. "You don't have to be anyone else. You can let go of needing to please everyone. When you do this, even the people who trained you to conform and the ones you most want to please will surprise you."

When companies tell women of color to "be authentic," this is part of the delusion construct we talked about in the previous chapter. Most companies want women of color to be *their* version of authentic, or "authentic within reason." But being authentic actually means meeting the world on your own terms by drawing on your own history and culture in order to thrive. Dr. Valerie Purdie-Greenaway, associate professor of psychology at Columbia University, says, "The critical distinction is agency and ownership. It is a source of power when you shed how you are being told to show up and instead find your authenticity on your own and use it to actualize your full potential."

In breaking through her delusions and finding agency, Lisa learned that she was enough, and this helped her realize she wanted to leave the corporate world and make it on her own without needing anyone's acceptance. She wanted to try and build her own business, even if it meant leaving her successful corporate path. In 2013 she left her high-powered job to start Gravitas, a successful clothing company. Now she makes no excuses for who she is and how she walks in the world—and it seems like her parents are also on board.

"My mom shocked me by incorporating my company for me," Lisa says. "She knew I was ready to be who I wanted to be."

To be honest, I was a little in awe of Lisa when I first met her. Discovering that her confidence was learned, and that anyone can eventually find their own power like she did, was thrilling. I became fixated on understanding how more of us, as women of color, could do the same. I want us all to feel unapologetic, grounded, and entitled to enjoy who we are and to feel power in the gifts we bring to this world.

In order to do this, we first need to become conscious of the messages we see and hear every day from many different sources—

our workplace, our cultures, society at large, and friends and family. Some of these messages are pervasive and damaging, especially when we're exposed to them again and again. They can strip us of our power unless we learn to shed the ones that do not serve us.

Typically, the women I met—including Lisa—went through a series of steps:

1. You bump up against a personal delusion.
2. You begin to question the messages and the indoctrination.
3. You work to shed how you think you must be.
4. You reclaim your own truth and begin to awaken to who you want to be.

More recently, many of the women I spoke to are looking at delusions because COVID-19 has made them reconsider the lives they have constructed and the work they want to do in the future. They are starting to ask new questions about how they want to work, how much they want to work, and how they want to leave their mark.

The delusions we are given usually uphold white male models and deny the magic in our uniqueness as individual women of color. These delusions consciously and unconsciously construct cages in our minds that can limit our imaginations and possibilities.

To move forward, we need to detach from those previously learned messages and permanently release false truths. Think about snakes as they evolve. They grow until a certain point, then shed their skin to make room for new growth. There's a lesson in that for all of us—we *must* constantly shed that which holds us back. Shedding is the active work that reprograms the limiting messages and stories. In a way, shedding allows us to come back to our true selves. Some of the ideas in this chapter are common delusions the WOC I met faced—but they're by no means a complete list of challenges you may encounter as you step into your power. Most of the women I met have had to shed one delusion or more to find happiness and power, especially at work.

THE SHAME OF INVISIBILITY AND DIFFERENCE

One of the biggest issues I heard repeated from the women I met is that they feel invisible. The world around us has contributed to this. Until maybe the last decade, we hadn't seen ourselves represented on TV, in magazines, or on the news. Growing up, most of us had to search hard for products made for our hair and skin tones. It was as if we didn't exist.

Not seeing ourselves represented, or seeing ourselves represented negatively through stereotypes, affects our psyches. Without having people around us who loudly and continuously tell us that we belong, many women of color believe that we don't. That feeling runs deep and causes confusion, pain, and shame.

Close your eyes and think of the word "executive." If you are like most people, the first image that comes to mind is not a person of color or a woman—much less both. We tend to envision executives as white men because that's how we've been conditioned. Whiteness is part of the "prototype" of leadership. Even as a former executive myself, I still have to consciously remind myself that I, too, fit the depiction of that word.

Our limited role models set us up for a life of constantly doubting our abilities, accomplishments, and opportunities. I remember walking into a client meeting two days after I made partner, trying to use my new title for the first time. I let them know I would be the partner in charge and explained my qualifications. My eyes were fixed firmly on my shoes. I heard the client ask my colleagues, "What's wrong with her? Is she *really* the one in charge?" My ability to make eye contact had deserted me. I was struggling with high-level imposter syndrome and a lack of bravado, despite being qualified for and having earned my new role. I overcame this as I ascended in my career, but it took a lot of conscious effort.

I want to come back to Lisa's story. "I spent most of my younger years pretending I was not Asian," she told me. From growing up Taiwanese in a relatively white town, to walking the halls of Yale, then reaching the junior partnership level at a prestigious consulting firm, where she represented some of the most desirable fashion brands on the planet, she always adapted. Lisa calls this Asian editing. In a way, it made her invisible. This type of editing builds a level of shame about being different that many of us bury to survive. The shame sneaks up on our inner power in unexpected ways and can cause us to second-guess ourselves.

While Lisa had every success she wanted, she still felt she was just getting by. To make matters worse, in her yearly review session, her boss said she had done a great job but that she lacked "gravitas." She didn't have the executive presence, the killer instinct, and the audaciousness for the next level of partnership. She wondered how that was possible, since she had given up so much and done what was expected to get where she was.

Over the next few months, shock and depression set in. Lisa was exhausted and decided to stop trying to make everyone else happy, including her mother, who had strong opinions about Lisa's life and career. Interestingly enough, by letting herself go, she found herself again. The process wasn't easy, but she stopped being embarrassed about her differences and stopped apologizing for what she wasn't.

Her story is similar to so many of the women of color I met. Once they stopped trying so hard to be someone else—usually as a result of having to navigate spaces where they are an "only"—they found ease, success, and personal power.

WESTERN AND WHITE EQUALS SUCCESS

As I met with more women, it became clear that the pressure to conform was part of a larger false narrative that whiteness and success are linked. This indoctrination runs so deeply in American culture that it is harder to shed than other delusions. Selah, an immigrant from Turkey, is a prime example of this. She explained that, early on, her parents taught her to blend in with her (white) surroundings. They believed that to succeed in the United States, her adoptive homeland, she would have to sound and act like the white majority.

Like Selah, many of us were taught overtly or subvertly that white women are more socially acceptable, and many of us have had to either build strong characters to validate ourselves or suffer from low self-esteem because of who we are. On top of not being seen, we don't get external validation in the ways that white women do with images and definitions of beauty. It can be hard to feel valued or beautiful when the absence of positive images tells many of us that we aren't acceptable as we are in society, and that we need to change and adapt to fit in and get ahead.

Jackie, a Black woman who works in the media and entertainment industry, grew up believing that being successful in the business world was less about developing her brain and more about conforming her hair. Among other things, her mom taught her to apply a chemical relaxer that straightened her hair every eight weeks. "If you want to find a job, if you want to find success, you must keep your hair together," her mom would tell her.

For Jackie's mother, being acceptable equaled acting and appearing white. This was the way she had found success, and she believed she was acting in Jackie's interest. But although Jackie credits her mom for preparing her well, "she also taught me *not* to be myself," she told me. "For a long time, I felt like I had to be like everyone else. The Jackie you saw in the office was not real—it was more like an office-appropriate mask I'd created for others', and mostly my white colleagues', comfort."

At a conference in 2019 I was on a panel with several internationally born participants when an Asian woman in the audience raised her hand. As she held the microphone, she immediately started crying. From watching our panel, she suddenly realized that she had been self-editing. Fighting tears, she explained that until she heard the other three Asian panelists speak about being senior leaders, she'd assumed they were also American-born. This audience member had convinced herself that being an immigrant with an accent disqualified her from the senior leadership ranks. Her view and perspective immediately changed when she had new role models with thicker accents than hers.

The experiences shared in these stories are common. They point to mindsets we need to shed. Whiteness is not a qualifier for success. Believing there is a causation not only places a high cost on the person who feels they must assimilate to succeed but also feeds the structure that upholds the status quo.

INDOCTRINATION OF STEREOTYPES

Pavi Dinamani, the cofounder of a video marketing business, MisFit Communications, originally created her business to help young millennials talk about the stigma of being unemployed. She found the pressure of losing her job especially hard to talk about around her immigrant family, so she started a YouTube channel to document her journey. Over time it grew into a platform that addressed many hard cultural discussions. She even featured her parents.

When Pavi first immigrated from Dubai to the US, she spent years as a chemical engineer in the oil and gas industry. She laughed when she shared that, in addition to having to transcend others' ideas that she was too quiet, deferential, and all the other stereotypes that come with being a woman from an Arab country, it took over a year for her to call her supervisors by their first names. She had been raised to call her elders "Aunty" and "Uncle." After a while, she transitioned to "Sir" and "Madam," before getting to "hey," and then, finally, shifting to a first-name basis. For years, her coworkers used this as fodder, treating her as their idea of a stereotypical immigrant. They underestimated her technical capabilities and didn't defer to her as a leader.

Many of the Asian women I spoke with shared that they count the number of times they speak in a meeting, making sure to speak up more than their peers, to subvert the stereotype that they are quiet and reserved. If you are an introvert, this can be especially hard. These women are trying to fight the idea that they have to act in support of the "model minority myth." Buck Gee, a former VP at Cisco, touched on this topic, saying that society has set Asian Americans up as good thinkers and doers, who lack the assertiveness, vision, and interpersonal skills to be successful organizational leaders. "The stereotype of the quiet, talented professional has led to the widespread assumption of an ill-suited business leader," says Gee.

Many of the Black women I met are constantly trying to manage perceptions of the "angry Black woman" at work. According to research done by Coqual, almost one in five Black professionals report being mischaracterized as "angry" at work. This affects how they dress and speak, and even the facial expressions they use. Ginny, a Black tech leader, is almost six feet tall and very conscious of how she appears when she walks in a room. She said that in recent years she's had to work on her posture because, for so long, she had slumped so she wouldn't appear in her full stature. On the one hand, she does think her height helps her to be seen as powerful in corporate settings, but on the other, it almost always makes her appear intimidating.

Latinx women I spoke with shared that they constantly fight against the stereotype of the emotional and feisty Latina. A few women I met even shared they had hired personal speech coaches to practice modulating their intonations. One woman said, by trying to adapt a more monotone voice, she was hoping that she could counter early career feedback that she was too volatile. Beyond her voice, she has now muted how she dresses and the colors she wears—and in order to blend in better, she won't let herself be heard speaking Spanish at work. She believes there is more reward in being "stoic than excitable." Some of the Latinx women I met shared they feel that, too often, people rush to simplify their ethnic identity when it is full of racial complexity. Some women shared that white bosses and colleagues will simply categorize them as Black or white instead of recognizing them as Latinx.

This rush to "stereotype" along one dimension of identity, in this case skin color, is another reason many women of color feel invisible. The complexity of our identities should not be simplified and condensed by our skin color. We can't shed the stereotypes others hold about us—that is work they must do on their own—but we can look at how much these ideas constrain us and make us edit how we walk in the world. It's a tough, and perhaps unfair, position to be in. But the more conscious we are of these biases, the more confidence and compassion we will have for ourselves. If we can shatter harmful stereotypes, we may also be able to pave a smoother path for the women of color to come.

SOME CULTURAL AND TRADITIONAL ROLES HOLD US BACK

Many women of color come from cultures where patriarchy and misogyny run deep. Across Asia and the Middle East, in Central and South America, and in parts of Africa, women's movements and rights can be limited and controlled. Some women from these cultures have struggled to find their voices, and their place in their family can be at odds with the power roles they play at work.

In my own culture, girls from certain parts of India have strong pressure to get married at thirteen and fourteen to reduce the financial burdens on their families. Child marriage, ostracization for being a widow, infanticide of girl babies, and dowries are very real parts of the history I come from.

These cultural norms often carry over outside of our ancestral countries, too. Growing up in America, arranged marriage was a big and contentious debate for my family. My mom has shared with me many times that she was considered a bad mom and a failure for "letting me" go for so long without getting married. As I went through my early career stages, it caused a lot of tension.

Indra Nooyi, the former CEO and chairperson of PepsiCo, and someone who is consistently ranked on the "world's most powerful women" lists, has also shared her story of arriving home to tell her mother that she'd just been named CEO. The minute she walked through the door, her mother asked her to fetch milk from the store like a good daughter. Her mother said, "You might be president of

PepsiCo. You might be on the board of directors. But when you enter this house, you're the wife, you're the daughter, you're the daughter-in-law, you're the mother . . . So leave that damned crown in the garage." Women are already assigned limitations based on gender, but for women of color who are the first, few, or only, cultural beliefs are layered on top of this, making it even more precarious when we show ambition and a desire for more.

A Chicana who has a senior role at a large consumer products company, Andie, told me her dad believed this job would stop her from finding a husband, and that she would be single forever. This was especially ironic, because she and her sisters gathered every evening when her father came home from his factory job. He would inevitably vent about some uncomfortable event, many of the stories having to do with his coworkers calling him racist names or slighting him. Andie recalls how her father felt powerless to do anything against them or report the events. He just kept quiet and internalized the pain. She somehow thought her dad would be happy when she reached senior levels of her company with more power to speak up and make decisions. Instead, he said her success would keep her from finding a husband, raising a family, and playing a more traditional role.

These examples speak to the duality of the worlds we straddle as women of color. Many of the women I met walked in two or more cultures: the world of white Corporate America vs. the different lives and cultures they lived at home. Whether this meant financially taking care of extended family, obeying more traditional roles, or living in very different racial and economic geographies from where they worked, the women I met made huge transitions between their work lives and their home lives.

As these stories have made clear, women of color from around the world struggle with cultural norms, whether in their family's country of origin or elsewhere. Because these mindsets are ingrained in us during our upbringing, they can be difficult to overcome. They seep into our work roles and the overall identities we create for ourselves. However, to be true to ourselves and reach our highest potential, we must recognize and shed these limiting societal constructs if they stand in our way.

OUR FAMILIES TEACH US HOW TO WORK

Ideas about work and its importance in our lives are often imposed on us by our families. Many of our parents, who were immigrants and outsiders, feared that attention would bring scrutiny, conflict, and danger, whether in their home country or in America. They often taught us not to "rock the boat"—to keep our head down and work hard, and only then would the American dream happen.

Black women have almost always been taught not only to work hard, but to be twice as good in order to get half as far. That means they must work four times as hard as their peers to be equal. Let me repeat that. *They must work four times as hard.* Because they have been raised to believe this, there is no space for error, and the only way to be taken seriously is to be impeccable. This has put a large burden on the Black women I've met, and it can lead to high levels of burnout. And working this hard often still won't make the ball bounce in your favor because, as we learned from Delusion 4 and the meritocracy myth, the system may not be set up equally.

We have been taught the great sacrifices our parents and ancestors made for us to be here, and many of us feel the weight of that history. This responsibility to community and family is often paramount for many women of color. It leads many of us to stay in workplaces or situations that don't serve us because we feel a sense of responsibility for holding the seat. Many WOC are also caught between the desire to push for change and go out on our own, and the lessons we've received that we need a stable and predictable income in order to survive. We are often taught to be grateful for being included—an implication to maintain the status quo instead of asking for more. This is in direct contradiction to the business world where we are told to "have a voice" and "speak our minds."

Like many women, I absorbed my Indian parents' fears. They had directly or indirectly pushed messages that traditional success was important. Where you worked and what you did mattered a lot. Worthiness came from "doing" and working hard. Sacrifices would be rewarded. Life wasn't about joy or fun. I needed to shed those messages and redefine what would make me happy because, like Lisa, although I was successful in my senior-level role, I wasn't really happy. Once I shed my parents' messages, I needed to rewrite my life in ways that may not have been familiar to the old me.

Most of the older women I spoke with were overtly taught they had no right to ask for or negotiate for what they deserved. They were taught to blend in and to not make waves. They all assumed that this had changed for younger women who are entering the workforce today. While I do think women entering the workforce today have less patience and louder voices on certain issues—sexual harassment as a result of #MeToo, for instance—many of the younger WOC I met were still struggling with how much they could push on the system. They were also uncomfortable with the idea that they might have to speak up and make waves in order to maintain their authenticity while moving ahead.

Many of the less-tenured women I met didn't see a long-term career in Corporate America. They were at their companies to learn skills and then move on to opportunities outside of large corporations. We will talk about this more in Chapter 8, but while they were in their corporate roles, they still felt the traditional pressure to prioritize keeping their paychecks, being liked, and being seen as good workers—all values they had been taught by their families.

ARE WE PASSING HELPFUL MESSAGES TO THE NEXT GENERATION?

Race, gender, and class are all ideas that we are taught. And, unfortunately, many of us have to teach the next generation some of these same messages and warnings. Donna, a recruiting leader who places executives on boards, is Black and married to a white man. They have three biracial children. The week after George Floyd's murder, her family decided to leave Boston to drive to the Cape. Much of the drive was on a two-lane road with a double line.

As they were driving, her six-year-old son asked her if it was against the law to cross the yellow line, and if a person would get in trouble if he or she was caught. Donna explained that they weren't supposed to, and yes, it is against the law, but she didn't think they would get in trouble because the road was empty. Her other son, a nine-year-old, turned to her with sarcasm that she described as masking fear and said, "Well, *we* may get in trouble. *We* may get put in jail even if Dad doesn't. It depends on who's driving, doesn't it, Mom?" She didn't know what to say. Her nine-year-old wasn't mistaken. We both cried as Donna recalled the story. She told me, "I wanted to dismiss what he said, but he isn't wrong."

As I was doing research for this book, I also met Deja. Her daughter is biracial and enrolled in a well-known ballet school in Los Angeles. A new teacher at the school suggested that Rachel wasn't good enough for the advanced class even though she had already been on the advanced track for a few years. When Deja confronted the teacher, asking why her daughter had been moved, the teacher said very matter-of-factly that "kids like Rachel are better off playing sports, not in ballet." It was clear to Deja what the teacher meant.

Deja's first inclination was to rip the woman's head off, but she has more poise than I do. She stopped and had an important inner conversation. Her daughter is ten. Deja knows the world is not always a kind place to biracial Black girls, and she decided to use the opportunity as a teaching moment about race. She told her daughter what happened and asked her what she wanted to do. Deja's daughter opted to stay in the less advanced class and has found great joy being one of the stars of the recitals with her friends and a teacher she likes.

Deja shared that the situation took years off her life because she was so afraid of harming or even muting her daughter. She knew it could and probably would serve as a huge checkpoint memory for her daughter, so Deja made sure to give her true agency.

Many of the women I met tell their children to "do more and do better" in order to compete and even survive in a white world. There's an expectation that we will dole out different messages than the ones we heard growing up because of the impact those messages had on us, but they often stay the same. In some cases, passing those challenging messages on is still a form of survival. It is why learning to shed messages that don't serve us as adults is so important if we want to change the world for future generations.

SO, HOW DO YOU SHED?

Shedding *should* be easy in concept. Phil, a medical intuit I know, suggested to me that shedding is parallel to breathing. We breathe air in, convert it by taking in the usable and helpful parts that work for us, then expel the rest. We do that unconsciously, and it is critical for our health and for life. He said shedding should be the same. It can be hard, and it's not innate like breathing, but it does become easier with practice. And no one is going to do it for us, so we must do it for ourselves.

Life challenges provide the perfect opportunities to shed messages that don't serve us. Lisa, for example, used the difficult feedback she got from her review to distinguish between what was sacred to her, and what she'd inherited that no longer served her. Shedding is not always voluntary or even intentional. Sometimes a situation at work—a bad boss or a lost promotion—or a personal "crisis" will call on you to shed if you are willing to accept the invitation. One woman shared that challenging events "can provide a portal to a new life," if we recognize them.

It can be hard to give up what we have been taught, and it can take outrage or sadness to be willing to look for something new—but the realization that we are living with these delusions gives us a chance to begin shedding.

Now that you consciously know that there are a number of personal delusions WOC face, you can begin to understand what you can do next. The following stages will help you move forward when you come up against these delusions.

BEGIN TO QUESTION THE MESSAGES AND THE INDOCTRINATION.
After you have identified the delusions you want to shed, listen to your own voice. For us to take our power back and claim our "real truth," we need to sift through all the messages we were given by society, our families, and the structures around us. To do this you must quiet outside noise, find silence, and take the space you need.

Some women do this work in therapy, others journal, and others create vision boards. Everyone's process is different. Rha, my business partner, says she needs to get quiet to do work like this. She meditates and keeps every Sunday open for time to reflect. The process for Rha is about integrating mind, body, and spirit, and she uses her faith and spirituality to do her shedding work.

This stage can take some time. Lisa took a year off after her corporate role and traveled. She wanted space away from her day-to-day life to figure out who she wanted to be and what she wanted to do next. She came up with her list of delusions she wanted to shed, as well as her vision for her new company, while sitting on a beach. She started writing on a piece of paper and the whole model just came together.

If journaling isn't your thing, just make sure you really sit with the incident that happened to you, or the patterns that keep coming up for you, to find ways to pull them apart and process them. The most important thing about this stage is giving yourself space and time. If you like to take walks or work out to clear your mind, try that.

If what is happening at work is opening your eyes to your own delusions, or you realize your company may not have your best interests at heart, try asking yourself the following:

1. What did you believe was going to happen?
2. What actually happened?
3. Why did it hurt you?
4. Why were you caught off guard?
5. What do you believe that sets you up to have different expectations?
6. What are you questioning about yourself, your life, or your beliefs as a result of the incident?

Answering these questions will usually result in some beliefs around delusional thinking. Let me say up front, this part of the process is not easy. When you start, it can feel like you don't know who you are or what you believe anymore. Many women of color are already so hard on ourselves, judging ourselves all the time, and as we shed, we have to come to terms with the ways in which we had to contribute to the system to get ahead. This can get confusing and very real. It is the hardest work.

WORK TO SHED HOW YOU THINK YOU MUST BE.

Now you know what you want to shed, but how do you actually do it? Sometimes, even when we know we must let go of something, it can be hard to follow through. There is always tension in what we need to release.

Barbara Adachi, a petite Japanese woman I've known for years, was one of the first WOC to serve in a senior role at a global professional services firm, where she had been a senior consulting executive. As someone who is "quintessentially quiet," she has had to step outside her comfort zone and push herself to speak up and be visible. Now retired, she sits on two corporate boards.

She comes from a cultural background where she wasn't always encouraged to be seen and heard—in fact, she was taught to keep quiet and not make any waves. She continues to work on shedding the messages that she should limit her voice by always coming prepared. Before every presentation or meeting, she thoroughly studies the material and prepares detailed notes. Then she "practices" her questions and any important points she wants to make in advance. While this approach wouldn't work for everyone, her preparedness gives her courage. "We are in the room for a reason," she says. "So, ultimately, we will need to shed our insecurities and speak up with confidence."

Barbara is one of the most accomplished women I have met and a personal role model to me. Even at her tenure, she is willing to openly share that, as a WOC in corporate spaces, she still has cultural challenges she is actively working to overcome.

I strongly relate to Barbara's desire to be ready. In 2019, I was taking a sabbatical from work to focus on my health, yet my type A personality was overwhelmed by the idea of "doing it right" and having a plan for my unstructured time. I was scared of leaving my role and everything I knew, so I met Tara-Nicholle Nelson for lunch on the suggestion of a friend. Tara is the founder and CEO of SoulTour, a personal growth company focused on transformation, and she helps entrepreneurs and companies unleash their creativity. She showed up in my life just when I needed her, and her work helped me move through the very difficult transition out of the life I knew and my corporate career.

She says most of us seek external validation and are living with symptoms of chronic hesitation, imposter syndrome, scarcity, and unclear direction. These are all driven by the same delusional belief system. I'm sharing one of the notes she emailed me because it stopped me in my tracks:

> Perform. So that you might have nice things and have nice things happen to you.
> Conform. Be a certain way. Stop being that other way.
> Produce. More. Now, even more. No days off.
> Perform. Conform. Produce. These are the dictates of culture.

Tara helped me see that I was living by these messages. I wanted to shed them, but I didn't know how. I eventually shed them by doing a regular daily practice of meditation, moving my body, writing, and affirmations. These practices helped create a new, quiet space so that I could hear my own inner voice. They worked because they created time for reflection and reprogramming. I followed this routine every day for a year, and it helped me rip away all the scaffolding of my life. As I reflected and wrote about the delusions I wanted to shed, it became easier for me to see my old behaviors and habits as they surfaced in the moment, and to make different choices.

I realized I kept trying to find ways to stay busy and "remain productive" because these were the behaviors I'd been rewarded for growing up and in my work life. But instead, I could just take a walk, read a book, or cook a meal. I enjoyed doing these things, but I hadn't always made space for them because they didn't make my "to-do list" or accomplish a stated goal. Small moves like these have led to bigger changes in my life over time. The soundtrack of delusions still plays in my head, but now I can identify each one. I am more intentional about how to stop them versus having to feed them in the moment.

RECLAIM YOUR OWN TRUTH AND BEGIN TO AWAKEN TO WHO YOU WANT TO BE.

Shedding our delusions will open doors we wouldn't have been able to walk through if we'd kept our heads down and operated inside a paradigm that wasn't built for us. Leadership has traditionally been narrowly defined and doesn't allow for styles outside the white male norm. We can shed the old delusions about what a CEO, professor, president, thought leader, founder, philanthropist, or pilot looks like and replace them with visions of these roles that actually look like us.

Once you begin to shed the old belief systems, you awaken to the possibilities available to you when you choose to live without delusion. KC, a member of nFormation, shared that she used to believe a common delusion: her worth and her value came from external validation and status markers. When I met her, she was working at a company focused on product innovation in the traditional energy space. Despite spending the last four years of her career growing her department, she didn't get the promotion she was expecting, and she began to question if she was good enough.

Over the next few weeks and months, KC went through a process of shedding that showed her it wasn't her worth that was in question, but the fact that the company wasn't valuing her expertise and the creativity she brought. After moving through a grieving stage, she realized that she was too out-of-the-box for her current company. She decided instead to focus on her unique skill set in a related but different industry: renewable energy.

She also realized she didn't want all the extra work that would have come with the new role, and she didn't even see herself in the company or in the South long term. She wanted to move west, where there was more opportunity and more diversity. The shedding process helped her get honest with herself—she had just been climbing the ladder with the delusion that a bigger title would somehow validate her value. The right answer for her was not at the same company but instead looking for a company where she could explore more innovative, green tech ideas.

••

Sometimes shedding means leaving behind everything you have been taught and instead following a completely new belief system or a call to happiness. For many WOC in the workplace, shedding is realizing that the systems around you are unfair, and some of the things you've been taught to strive for may not bring you happiness, and may even disrupt the current illusions you are holding around contentment. Sometimes, shedding requires a leap of faith. Every woman I met who took this leap found bigger opportunities after she jumped. Some took this leap in their existing jobs. Others did this by moving on. But ultimately, understanding and releasing their personal delusions brought a sense of freedom and liberation.

Shedding is about the person you haven't yet let yourself be. Once you shed what doesn't serve you, you'll make space for your own truth, be able to write your own narratives, and decide what you want to carry forward.

CHAPTER 3

CARRYING WISDOM THAT FEEDS US

The idea that certain traditions and beliefs should be carried forward is parallel to the shedding process. You can't possibly shed everything you have been taught, nor should you want to. Part of our power as WOC comes from our origins. They give us a sense of grounding. When institutions and structures don't validate us, we need to remind ourselves of who we are. In a world where we are told to assimilate, we need to take pride in where we come from.

I have met many inspiring women through writing this book, but one stands out as the embodiment of inspiration. Rani's joy, heart, and radiant energy immediately capture you when you meet her. She has had to shed more than most of us will ever imagine, but she's also learned what to carry with her.

Rani was born with cerebral palsy. In the rural village in India where her parents were raised, people believed that a child born with disabilities is a result of karma and sins from a past life. There is great shame brought to a family when a baby is born this way. Instead of carrying the trauma of her birth, Rani's parents told her that people given challenges were special in the eyes of God. So special, in fact, that if God hadn't given her some extra challenge, she might be too special and too godly for this earth. She was raised with the saying "Many things are possible, if you don't know they are impossible." To give Rani the best opportunities, they immigrated to the US for a new life. Unfortunately, though—and this applies to many of us women of color—as Rani grew older, the "brainwashing and hypnosis" began.

Her advisors told her to be realistic and "reasonable" about what was possible for her and were frustrated and angry when she wasn't. They wanted her to give up her dreams of having a family because of the physical challenges to her body, to let go of ideas on overseas travel because of accessibility issues, to get her degree online instead of navigating a college campus, and to stop dreaming of working for a global firm in a traditional office role. They wanted her to acknowledge *their* limitations for her. Rani chose differently, deciding to believe that anything was possible.

Rani believes the stories you tell yourself and the stories you let others tell you determine the quality of your life. She chose to bet on herself and, as a result, she spent a year working with Mother Teresa in India, earned a graduate degree in international business and corporate anthropology, has a family with four children of her own, and is head of employee advocacy at Adobe, where she leads with her head *and* her heart. Rani has learned to embrace what makes her special and unique and has cultivated a level of power and inner strength that I hope more of us can attain.

Drawing from our challenges is our superpower as women of

color. Not being seen and heard has made us wise, clear, and ready to step into action. One of my teachers says that real power is a form of alchemy. It is being able to take difficult situations and turn them into wisdom for others.

There is an Indigenous story I've heard that also reinforces Rani's message. A grandfather speaking to his grandson said, "A fight is going on inside me. It is a terrible fight, and it is between two wolves. One is evil—he is anger, envy, sorrow, regret, greed, arrogance, self-pity, guilt, resentment, inferiority, lies, false pride, superiority, and ego. The other is good—he is joy, peace, love, hope, serenity, humility, kindness, benevolence, empathy, generosity, truth, compassion, and faith. The same fight is going on inside you—and inside every other person, too." The grandson thought about it for a minute and then asked his grandfather, "Which wolf will win?" The grandfather simply replied, "The one you feed." I remember this story often and try to use this wisdom in my own life. We need to starve the "evil wolf" and shed the harmful messages. And we need to feed the "good wolf" and carry forth the messages that serve us, like Rani did.

I want us to root in order to rise. Think of a plant growing, the stem elongating, the leaves opening up as it stretches toward the sun. Think of all those time-elapsed clips you've probably seen where plants are shown growing quickly, up and out. Now think of the part of the plant we don't see, the roots below the soil that we tend to forget, the real source of the plant's strength. I want us to root so we can carry the wisdom of our origins with us.

Whether we have the privilege of knowing our lineage or are still filling in the pieces of our exact history, there is so much power in "naming and reclaiming" our history. Doing so is a powerful way to rebuild identity.

CARRYING THE ANCIENT WISDOM OF OUR CULTURES

Carrying wisdom may be the most important solution for what ails our world right now. Wisdom is the opposite of delusion. Delusions are rules that aren't based in history or fact but are self-created (mostly by white men) and set in place to control women's behaviors, their narratives, and their power. Wisdom is not about control; it can stand the test of time. It arises from experiences that tell us how to live in a way that is concurrent with our innate values.

Regional Chief Kluane Adamek is a citizen of the Kluane First Nation and honors the matriarchs who welcomed her to the Dakl'aweidi (Killer Whale) Clan. She has served as the Assembly of First Nations Yukon regional chief since January 2018 and is the first woman to do so. She has held various roles in the corporate and private sectors and in government. When I watched her TED talk, "The legacy of matriarchs in the Yukon First Nations," I reached out to speak with her. She says some people see wealth as power. But she is reminded of the elders and their teachings. Elders have stories to tell and "having stories is wealth."

She explained that potlatch is part of her, and many First Nations cultures. Feasting and celebrations take place, where gifts and possessions are given away to guests. "How much you give, not how much you have for yourself, is how wealth can be considered," she said. Knowing who you are and where you come from (when you can) is power. "Power in knowing and being grounded in that." She explains that drumming and singing are used in her culture because they bring who you are to the surface through vibration. There is power in the self, and it's in our being. For the chief, "women have skill and power because of who they are and their story."

In Clarissa Pinkola Estés's book *Women Who Run with the Wolves: Myths and Stories of the Wild Woman Archetype*, she explains that rituals and traditions, especially related to women, were often not written down. Because women weren't always taught to read or write, their personal histories and stories were lost. As a result, we have to consciously carry the wisdom we do have, or systemic thinking will replace it with delusions that serve the system rather than us.

Inherited cultural wisdom can take on many different forms, but much of it is being stripped away by Westernized societal norms. For example, Western medicine has superseded the cultural wisdom passed down from grandmothers and great-grandmothers on the power of plants to heal. One South American woman described her grandmother knowing how to treat every ailment in her family from her pantry. Big farms and genetically modified food have largely taken the place of our grandmothers' wisdom about how certain recipes can nourish the body.

This is also the case with beauty, where commercialized cosmetics and plastic surgery have overshadowed cultural remedies that amplify natural beauty. One African woman I met described the many tips her grandmother taught her, including using natural oils and seeds to nourish her skin and keep it looking supple into her sixties.

We need to keep holding on to the valuable parts of our ancestorial wisdom even if modern society has contorted how we should see them. Asian women have often been deemed less powerful and strong for being quiet and reserved, and this interpretation feeds into the model minority stereotype instead of seeing the value and positivity in these traits. One Japanese woman shared that, rather than valuing individualism over the good of the community, her great-grandparents taught her that strength and success came from being proud but humble, staying centered on inner values, and being an integral part of a community. Many of the Asian and Middle Eastern women I interviewed said their cultures value honor and grace over competition and backbiting. Carrying the wisdom from our cultures gives us the tools to question the predominant delusions in corporate culture so that they cannot take hold.

CARRYING THAT WE ARE UNIQUE

As women of color, we have a lived experience of being double outsiders, and this gives us a unique perspective on American society and business. During my research, I met other women who are studying aspects of the first, the few, and the only. Angela Chee, a communication coach and former TV news anchor, has a podcast called *The Power of the Only*. As we compared notes on the women we both met, it became clear that many of us have been toughened up in some way, and those experiences made us stronger and defined how we show up in the world. As we said earlier, our obstacles became our power.

Although we have taken unique paths full of determination and grit to get where we are, we are expected to adapt to the business culture around us when we get there. The system doesn't recognize that we as women of color are different, and yet we know deep inside we have to maintain and even harness our differences to be our best. The richness of our diversity vanishes because we need to be like everyone else. Because of this pressure to blend in, many of our new perspectives and views that propel innovation and business success are dismissed. This might be the biggest loss to organizations.

Angela believes the women of color who are most powerful have seen their strength in both worlds and have charted a path forward by being able to navigate and bring the best of each to the other. "We see 360 degrees," she says, believing we have a greater depth of understanding and empathy because of our challenges. "We can see bridges that others can't," she continues, "and we can make sure, in ways no one else can, that voices don't get lost." She believes that together we can create what's next.

CARRYING EMPATHY, EMOTION, AND INNER KNOWING

In the corporate world, logic and fact trump almost all other ways of operating. This doesn't always leave room for empathy, collaboration, and diverse paths to decision-making. As we walk in corporate halls, we've had to mask our own intuition and feed the delusion that there is only one way to lead.

But many WOC say they lead best from their sense of inner knowing and gut feelings. My big decisions were based less on fact, and more on an inner feeling about what I had to do. One of the women I met calls this "Whole Heart Leadership Style" and suggests that WOC are taught to push this deep wisdom aside in preference for logic and facts. Dr. Valerie Rein, author of *Patriarchy Stress Disorder: The Invisible Inner Barrier to Women's Happiness and Fulfillment*, validates the irony suggesting many successful male founders and executives say they rely on their intuition when making important decisions. Yet many women are ridiculed for using the power of our inner knowing.

In fact, McKinsey published an article in 2020 talking about the very qualities we need to lead in a crisis, like awareness, vulnerability, empathy, and compassion—suggesting that to move forward in the world, the characteristics and ways of leading that got us to this point will not get us where we need to go. WOC cultures did better than white Westernized culture to place importance on the characteristics McKinsey points to, including creativity, spirituality, healing, justice, morality, balance, deep listening, and passive resistance. We need to carry these forward as we lead in the future.

Intuition and empathy can be cultivated by allowing ourselves to explore what some might label as "woo-woo," but self-awareness and self-discovery are key ingredients to strong decision-making. Ginny, the Black tech executive I mentioned in the last chapter, says her father's death was the event that triggered her to move beyond systems thinking and begin her path of self-discovery, and this deliberate spiritual journey saved her. Today, Ginny leads with self-awareness and wholeness. While she had always led with her intuition, she hadn't always trusted it. "Now I can draw upon years of generational wisdom," she says. "I can draw upon what I'm seeing, hearing, and feeling. It's like a spidey sense. So many women I know do it instinctively, even if we never talk about it."

Sometimes the skills of intuition, empathy, awareness, and compassion are developed in response to having to deal with delusions. As I mentioned before, Ginny is six feet tall and had been told for years she was "intimidating." In order to appease that perception, she chose to find ways to make herself more accessible. Knowing her height wasn't going to change, she developed her empathy quickly and deeply. Now she is driven by the belief that if we meet people with empathy and compassion, good things will happen. This practice helps her stay in touch with the people she leads.

Ginny admits that at her age and stage, she still has to pull out the programming and the disempowering messages. But, she says,

"This is the time of the divine feminine, and we are unstoppable. Until now, we have been told we weren't as good as those around us. We have had to assimilate to participate. We have had to survey the lay of the land and learn to anticipate. And we have had to learn what to do when we are gaslighted."

Part of how we carry our emotional intelligence is knowing when we need to get angry. Despite being taught that anger needs to be covered over, diffused, and shed, if we are going to rise in power, we are going to have to embrace and cultivate our anger when behavior and decision-making are unjust. I am not talking about violence; I am talking about allowing ourselves to carry the whole slate of emotions. This means ending denial and moving into defiance. In *Rage Becomes Her: The Power of Women's Anger*, Soraya Chemaly says, "Anger is an assertion of rights and worth." When we get comfortable with anger and give it permission to show itself, we can unleash the power that comes with it.

Any time you feel an emotion, it is part of your inner knowing, your inner moral compass. It can help you know when something isn't right, when someone has said or done something that doesn't work for you. The most important thing we can relearn and carry as women of color is how to use that inner knowing, our intuition, and feel proud of that.

CARRYING A DESIRE TO CONTRIBUTE

When the dominant culture was enslaving or controlling them, our ancestors learned how to survive as a community. Our very existence represents their resistance and courage. Today, we can carry forward the deep sense of what has been sacrificed for us. We can feel gratitude and humility as we walk and work in the world. The desire to contribute is gleaned from being part of that long lineage of ancestors who gave us a better life. There isn't always a place for these stories in modern life, but we carry them with us whenever we give back to others.

Most of the women I meet carry a desire to give back and contribute. They have a strong sense of family and community and a belief that we must lift those who come up around and behind us. Cara, a Black woman who works in community organizing in New York, feels strongly that women of color have a sense of deep respect and introspection for where they come from and a deep sense of responsibility for giving forward even better than they receive. "That's unique to us," she said. "I don't think white men and women have that in the same ways we do." She believes this comes from the idea of the collective and how many WOC come from communities and cultures that emphasize group and community success, not just personal wealth and ascension.

Sofiya, a young millennial, started a company to help others while also building a viable business. She moved to the US from India when she was five years old, leaving behind extended family, traditions, and familiar ways of life. As someone who grew up straddling cultures, she often felt a sense of displacement and a desire to belong. She believes that we, as immigrants and women of color, don't talk enough about how assimilation can lead to erasure and nullification of identity.

Over time, she began to realize she could honor and reclaim her Indian heritage and celebrate her American upbringing. She left her digital media career in 2020 to follow a greater purpose, and started This Same Sky, a company that supports fair-trade artisans in India by selling their products around the world. Through her new endeavor she is following her artistic passions and connecting back to her heritage and her history.

CARRYING THE STRENGTH OF PERSEVERANCE

Our power comes from a lineage of perseverance and continuing to push forward and even change the rules in the face of obstacles. Vice President Kamala Harris has talked about how she developed determination and endurance. "You know, I have in my career been told many times, 'It's not your time. It's not your turn.'" She then added, "Let me just tell you, I eat 'no' for breakfast, so I would recommend the same. It's a hearty breakfast." Stacey Abrams has also transformed obstacles into fuel. She turned an unsuccessful bid to governorship into a mandate to rewrite and rethink the voter sign-up process. Some would say she single-handedly changed the outcome of the 2020 presidential race. She didn't see the loss as a reason to back down, accept defeat, or stay silent.

The truth is, as WOC—and especially Black women—we will be told "no" more than any other group. We will be told it can't be done the way we envision. We will be cut at the knees. But if Stacey serves as any example, sometimes "no" means it's time to transcend the old constructs and do something we hadn't planned. The result is that, unfortunately, we need to have a higher level of resilience and strength than most.

Noni Allwood, a retired chief inclusion officer, spent years at Cisco and now sits on boards and helps conduct research on WOC topics. We share a deep interest in helping WOC understand the bias and the challenges that stunt us in the workplace. In her own life, she had to shed everything she knew and take a leap of faith that led her to a career most of us would look upon with awe.

At a time when few women were doctors, Noni's mom was a well-known medical doctor in El Salvador. Noni's mom differed from the other children's mothers because of her demanding career and her visible role outside the home, and this helped Noni believe that women could break boundaries.

Noni went to engineering school, one of only six women to graduate. She got married at twenty-one, had a daughter by twenty-five, and got divorced shortly after, but there was no place for a divorced woman in El Salvador. Instead of suffering through the marginalization she might experience, she carried her mother's example of charting new waters. Early in her career, Noni met the founder of a small software company in the United States and kept his contact information. She had impressed him with her expertise and industry knowledge. After her divorce, Noni found the courage to cold-call him and ask him for a job, knowing in advance it would be a steep road. He hired her and helped her move to the USA. She packed up her daughter and emigrated in the middle of El Salvador's civil war.

CARRYING OUR COMPLEXITIES

As women of color, our identities are intricate and our histories are complex. Noni says women of color's perspectives and understanding of the world around us are shaped by where we were born and how we got to the States. She referred to a framework outlined in the book *Hispanic Game* by Carlos de León de la Riva, which describes first-generation immigrants as "transmitters of culture," since they bring with them traditions, stories, rituals, and beliefs from their home country. Meanwhile, second-generation children of immigrants are "interpreters of culture," often caught in the middle of nowhere because they navigate two cultures. In either case, people treat you like an outsider.

The complexity of bias and stereotypes differs by culture, Noni says, and these stereotypes can impact the roles people think we can and should play. For instance, the "model minority myth" can make it hard for Asian women to be hired into roles where they lead others. Noni believes that for Latinas, it's even more complicated because they are multiracial, so their histories are mixed and nuanced. Some Latinas can pass for white, Noni says, and if they take white names, they may have an easier time navigating Corporate America as a result, but it becomes more challenging for them to relate to one another and determine where they belong.

Many WOC come from immigrant families or descendants of enslaved people and carry with them the ability to set out in a new direction and stretch boundaries. They have left everything behind or been taken from everything they know to start over. This includes their countries, languages, customs, and in many cases their power. By immigrating, some master resiliency, creativity, adaptability, and critical thinking. This makes them incredible assets to any organization, if only more were open to seeing past the stereotypes generally directed toward immigrants. And even though many immigrants and other WOC may be seen as quiet, they are true risk-takers for having started over. They carry a quiet strength. And although difficult to fully put into a few words, the descendants of enslaved people have had to learn to survive and find joy.

There are challenges that are unique to each of our identities as WOC. Today Black women are being asked to carry even heavier roles than they played before, addressing racism in workplaces. They often fear for their lives and their family's lives and have to leave that fear outside corporate doors. Latinx women are dealing with the impact of a political climate that is constantly hitting them with issues of immigration, border control, and an increase in overt racism. One of the Latinas I spoke with said, "The challenge for many Latinas is that the USA only sees us as poor women with children in their arms, asking for help." She says many people see Latinas as "uneducated, poor, and in misery," not as net contributors. Asian women are having to address racism and violence in the aftermath of COVID-19 being dubbed the "China Virus," and many Middle Eastern women are still trying to sort through the discriminatory impacts of 9/11. Indigenous women have been silenced and made invisible to much of the conversation.

Beyond race, there are layers to our complexities, including class, sexual orientation, disability, and gender identity. Stereotyping and discrimination push us to create an even stronger sense of who we are and to remember where we come from.

CARRYING THAT WE ARE ENOUGH

We recently did a photo shoot for nFormation, the company I cofounded. It was hard to find ten women diverse in age, body type, race, ethnicity, and industry. When I found the right mix, I was elated. But three days before the photo shoot, one by one, seven out of the ten women asked if they could back out.

Some of them wondered if they were senior enough and had enough tenure to tell their story. Others were worried that because they were biracial, they were not Latina enough or Black enough. They asked me if the other women would think they were too white or too blond. These feelings were palpable and profound. Here I was asking the women to come as they were, and they weren't sure if that was enough. Looking back, it makes so much sense. We have had a lifetime of people telling us we aren't enough.

When Ginny, who I mentioned earlier, talked about how companies consistently give us messages that we are not enough, she said, "I couldn't rely on my workplaces. I had to know I am worthy. Companies like having us on show, but we are tolerated, not included. We must validate ourselves and what we know. We need to love ourselves because others around us, others who sit in the structures and carry the delusions, won't love us." While what Ginny has described is a reality many of us have faced or are currently facing, we are reminded that we need to stop feeding this idea and start to carry that we *are* enough.

Some women, especially Black women, were told they were enough by their mothers or grandmothers, and these women almost always started with these stories when I asked what they carried. But the outside world hasn't given them this message to carry. And most of the other women I met didn't receive this message from home or in the world. When you don't believe you are enough, when you live in structures that ask you to give up more and more of who you are, the results can be devastating.

Confidence, clarity, and alignment come from knowing you are enough, and remembering that is a self-fulfilling prophecy. The more you know you are enough, the more the world will be forced to accept you. Knowing you are enough allows you to be unapologetic, audacious, brave, and bold. And this brings you closer to the magic of being able to create what you really want.

RECLAIMING OUR FULL POWER

Reclaiming who we are and what we value is how we step into our full power. This isn't a linear process. The most important part of becoming who we are meant to be is remembering where we came from and reclaiming all the lost parts of ourselves. We get to decide what empowers us and what we carry forward. We come from cultures of wisdom, and we can go back and pick up the messages we have been asked to leave behind. Similar to shedding, some women discover what they need to carry by listening to their inner wisdom through journaling, meditating, walking in nature, and sitting in silence. Some women find their history by exploring family trees, listening to music, dancing, looking at old family photos, cooking special recipes, or retelling stories that have been passed on for generations. Others send away their DNA to learn more about their origins. The real magic is finding what works for you. What will help you connect to your past? What will help you carry your power?

Tara, from Chapter 2, explains that rituals are one way we get ready for our work and honor who we are. She sees rituals as a sacred pause. Rituals, she explains, are "like a querencia." The querencia is a space in the ring of a bullfight. It is both a geographic place and a place in the bull's psyche where, even among the madness and the fear and the rage, the bull can find a temporary refuge. If a bull stops reacting to incoming stimuli and reaches back into his natural strength and power, smart bullfighters know to get out of the way. Tara says that a bull that finds its querencia becomes unstoppable and will run through anything in its path. For women of color, she says pausing to take part in a ritual can ground us in our innate power and knowing who we are, so that we, too, can become unstoppable.

Many WOC cultures practice rituals to remember their personal querencia, their connection to the earth or spirituality, and to focus on something bigger than themselves. Indigenous women have done rituals to remember their connection to the land and their ancestors.

Tara advises on the importance of implementing daily rituals, such as five deep breaths before drinking your coffee in the morning or writing in a gratitude journal every night. More people are turning to smaller rituals to remind themselves where they sit in a larger world, and to honor where they come from and what impact they want to make. Companies have even started to contact Tara, asking her to help them create and institute rituals in their daily rhythms and work culture. For WOC, ritual is a way to remind ourselves that we are special and have value.

Other women I met actively ask their ancestors for help through ceremony and prayer. As WOC, we come from strong histories. Histories that have suffered and histories that are sometimes glossed over and hidden to make the white majority feel more comfortable.

When I was doing research, I reached out to co-conspirator Elizabeth Lesser, cofounder of the Omega Institute and author of *Cassandra Speaks: When Women Are the Storytellers, the Human Story Changes*, a book about patriarchy, systems, and power. Through her work, she has organized several women's gatherings and written about women and feminism, and she shared a story with me that I want to share with you:

Years ago, Maya Angelou came to speak at an Omega women's conference. She walked in with dramatic flair, clutching her chest and touching her brow, saying, "This room is so crowded, I can hardly see." There were maybe four hundred women in the room. She paused for dramatic effect. Then she said, "Oh, I see thousands. No, millions."

Elizabeth shared that Maya Angelou then grew quiet and spoke to just the WOC, who were not a large number in the room. "Your mothers and grandmothers, your great-grandmothers' mothers are in this room. Your ancestors are here, and they are all so impatient for you to step into your glory."

Afterward, Maya Angelou led the group through an exercise about reclaiming power, which Elizabeth briefly shared with me. We need to be asking ourselves, "Who is with us? Where are our ancestors? How do we connect and speak with them? And what are they begging us to carry in order to shed our shame?"

So often WOC have been told to erase parts of themselves to fit in, to move forward, to survive. Sometimes we even feel we have to do it to ourselves. But Elizabeth shared that in doing so, we are cutting off our chi, our life force. Your lineage is in you. It's in your DNA. And if you lean into it and carry its wisdom, it will support you as you move about the world.

••

I remember being in a therapy session almost fifteen years ago, and my therapist Carolyn knew I was obsessed with dogs. She explained that I was still very anxious in my energy and demeanor after a recent incident and told me to watch my dogs in order to understand how to get my innate power back.

She said there is a power when they are about to attack or pounce on something, and she described it as power that is forward leaning. It is aggressive, and it is in your face.

There is a different power when they are comfortable, relaxed, and in their full being. They are on their back haunches, always ready to pounce if necessary, but settled.

That power is a healthier power. Finding our true power comes when we see through the delusions in the structures around us, shed what we are given that does not serve us, and carry forth who we truly are.

PART II

FEEL YOUR POWER

CHAPTER 4

WHEN YOUR MIND AND BODY SPEAK

"Your job is killing you," my doctor blurted out after my fourth visit with her. Dr. Heather had run a battery of tests, at least thirty to forty, and nothing had come back conclusive. She agreed something was off, but there were many varied, unexplained symptoms, rather than a big smoking gun.

I had adrenal fatigue and leaky gut, had recently been hit with shingles, and had *at least* fifteen other symptoms including hives, rashes, joint pain, headaches, constant pins and needles as well as numbness in my arms and legs, and extreme fatigue that no one could explain. I felt like I was dying slowly, almost as if someone were draining my battery, and I couldn't fully recharge it anymore.

Dr. Heather closed her file and looked at me. I was sitting on the examination table, and she was sitting at her desk, her chair swiveled toward me. My purse and my Rollaboard were tucked neatly in the corner. I had come from the airport to see her while I was on a business trip.

"We can keep going with more tests," she said. "Or I can be honest with you. I think you are living to work, and that is a recipe for disaster." I caught her looking at my suitcase out of the corner of her eye. "You're so stressed, your body is so stressed, and I think your lifestyle is the problem."

Then she started asking me questions that I had not asked myself in an awfully long time:

What makes you happy?
What gives you joy?
When do you feel most alive?
What would you do if you quit?
Would you be okay with doing nothing?
Do you see you are valuable for just being you, not because of what you do?

I choked back tears as I listened to her, willing myself not to let one tear drop. Emotional, not because what she said wasn't right, but because it landed so clearly. What she was saying I already knew deep inside; I just couldn't fully acknowledge it for fear of totally falling apart. I knew that once I admitted how tired I was, I might be breaking down the dam, and I might not be able to rebuild it.

I had been trained to think of myself as an elite athlete. I had made a career of ignoring how my body felt. I was resilient. I performed. I was a productivity junkie, and it made me feel good to get things done, traveling to three cities in one week, sometimes skipping meals and not sleeping, just to pull off results I was rewarded for. My superpower was outworking everyone around me. I wasn't the smartest and I wasn't the best, I just lived what so many little

Black girls and daughters of immigrants are told: *work harder to get ahead*. It was in my brain, and it was in my bones.

If I let myself acknowledge what Dr. Heather said was true, I was afraid I would have to do something drastic and change everything, including my job, and I wasn't sure I was ready. I did live to work. It was my entire identity. I was the A+ student for the perform, conform, and produce behavior Tara talked about in Chapter 2, and it had finally caught up with me. This was my fourteenth doctor. All the others had told me it was just part of the aging process. This doctor was telling me what I already knew but did not know how to change. She was telling me the delusions had caught up with me.

Dr. Heather was a doctor I saw when I traveled to San Diego weekly to visit my client. I was there more than I was home. When I left her office and went to my hotel room, I called a woman I worked with, June, who had seen me in tears in the office a few weeks before and had offered help. I took her up on her offer to call her dad, an infectious disease PhD and a rheumatologist, a doctor who figures out what is wrong with you when no one else can (like Dr. House from the eponymous television show). After a few months and another array of tests, he found I had late-stage Lyme disease. I finally had a diagnosis, and that diagnosis has led to almost four years of therapies and treatments.

During my journey I met healers, some of whom are WOC and who work specifically with WOC. Many of these healers have studied ancient Indian, South American, Indigenous, or Eastern modes of healing. Queen Afua, an author, holistic health practitioner, and wellness coach, teaches communities of color to go back to their roots and rediscover natural ways of eating and healing the body. She focuses on ancient wisdom and techniques from Africa to detox and rebuild, and she believes families and communities need to heal themselves in simpler, more traditional ways.

I have come to believe the Lyme was a gift, maybe a message from the universe, that I wasn't on the right path. Lyme caused me to really examine who I am, what I am, and what is important to me—and it helped me see I have a different superpower than I thought. My actual superpower, the one I have when I push past my own internalized delusions, is taking a setback or a challenge and using it as fuel to change my life. This illness sent me a message that I needed to stop and listen to my body.

In the type of career where most of us are financially and emotionally rewarded for ignoring our bodies, my body had finally had enough and was throwing a full-blown tantrum. The universe was unplugging me. It was time to pause. What I have learned, and am still learning, is that ignoring what we know and what our bodies know will catch up with us. My Lyme diagnosis made clear for me that I wasn't living my truth. I was out of alignment with myself and my power.

THE PRICE OUR MINDS AND BODIES PAY

Hands down, the biggest surprise in my research has been a trend I've seen among the most accomplished women of color: we are all exhausted, suffering, and yearning to find a way to keep our plates spinning. Most women of color believe they must work harder to prove themselves. The truth is delusions, microaggressions, and racism build up over time. As we try to push through and prove ourselves, we internalize them, despite and maybe even *because* we know the system can be racist. That's what often gets us sick. The World Health Organization says that discrimination is a health risk, in some cases even more impactful than lifestyle and healthcare choices. Scholars found that perceived racism and discrimination are linked to increased risk for hypertension, infectious illnesses, and a lifetime of physical diseases.

When women of color stuff down the pain and frustration that come from unfair practices at work, it sets them up for a huge struggle, pushing them to work too hard to live up to their employer's expectations, along with their own. In fact, there is a growing opinion that beyond high levels of stress, WOC are masking significant trauma at work. This trauma comes from feeling on guard because of racism, never believing we fully belong anywhere, and not feeling a sense of "psychological safety." The cumulative experiences of confronting race-related stress, emotional abuse, and the psychological trauma of racism can lead to what Anderson J. Franklin et al. call "the invisibility syndrome." In these situations WOC don't feel their talents and even identity are seen because of the dominant attitudes and stereotypes, and this leads to debilitating symptoms. These feelings cause an almost constant fight-or-flight reaction in many women. It's finally time for us to stop and listen.

Two out of three women I met had some type of chronic condition. For some, the issue manifests as a series of skin rashes, adrenal fatigue, or fertility issues. For others, it shows up as a mysterious stomach disorder, perhaps the result of what some women call "ingesting ideologies." Almost all the women who were ill or stressed had one of these four complaints, and medical studies support these findings. But it can show up in myriad ways, including chronic fatigue, insomnia, depression, anxiety, and more.

Imani is a senior technology leader and data security expert who's well-known in her field. I want to share her story first because, unlike other women I spoke to, she focused on the mental health aspects of being a WOC at work. Imani's challenges manifested as depression and anxiety versus other physical symptoms I saw in many of the women I met. She was in a serious car accident that required extended recovery time, and after a few months her company questioned the extent of her injuries, even after she presented the required medical backup from her doctors. Having her integrity questioned made Imani look at the delusions she held about her company. She now believes her company didn't believe her because she is a woman of color. "We aren't aware what's happening is actually racism," Imani told me. Delusion 4, the myth of meritocracy, tells most of us if we just work hard, we will be successful. But that's wrong, Imani says. First, we blame ourselves, then we lose confidence, and finally our self-esteem erodes. Over time, our physical health deteriorates, and our mental health gets tested.

Imani believes when we try to leverage company processes to address the issues as they become intolerable, we come up against processes designed to isolate and silence. When we dare question what is happening, we realize we are dispensable. That cycle shocks most of us physically and mentally. "In short," she says, "addressing the human impact of covert, overt, and systemic racism literally makes minorities ill."

As high-performing women of color, we need new ways to combat the stress and the trauma. Some of us may hold positions of power on the outside, but physically and emotionally we feel drained and powerless. Fighting a stress-manifested illness often means our body is asking for a pause. That signal can get so loud, the only thing to do is stop and listen.

The reason we get sick is that we are facing an invasion of our boundaries. In response, we need to strengthen and shore up our protection. Having healthy boundaries isn't something most people are taught. It is something many of us must learn over time. But understanding our limits means knowing what we need and being able to ask for it.

Dr. Valerie Rein has studied and written about these issues for women in the workplace in her book *Patriarchy Stress Disorder.* She says that when women believe tolerating hostile environments is necessary for success under patriarchy, we adapt to chronically high stress levels, workaholism, and "unconsciously shaping our ways of thinking, acting, and being to be more like men." These issues, she explains, go hand in hand with problems relating to sleep, weight, adrenals, and the thyroid.

When women find her work, Dr. Valerie says it usually results in an awakening of sorts. Most women who come to her have started their careers young and "done it all, achieved it all." They have climbed to the top of their industries and believed, once they got to a certain level, people would listen to them. But when they come to Dr. Valerie, they feel that they are "the only woman, the only woman of color at the table, and the only person that nobody is listening to."

She explains that women on this path have sacrificed and put much of their personality on the sideline, to the extent that they no longer remember who they are. They are awakening to delusions and realizing they are in a structure that was not designed for them. They shut down large parts of themselves in order to rise, chasing goals that have been prescribed for them, whether it's the next promotion or the next level. By the time Dr. Valerie meets them, these women are already seeing impacts on their health related to the need for women to comply with the rules around them, to contort and not be authentic to who they are. "This powerful woman of color, for whom the system was never created, never feels safe," she says. "She keeps reading books and trying to fix herself, but it's not her. It has nothing to do with her."

It is common for healers to say, "Disease means dis-ease, when your body is out of ease." My therapist once told me that so much of what is happening to women is not being comfortable in our skin. Our bodies are therefore literally eating us from the inside out. Rha suggests that men are taught to act out, but women, especially women of color, are taught to "act in" and internalize the pain and shame.

Blanca, a senior executive in her early forties working in the manufacturing industry, fought to reach the highest level a woman, and a woman of color, had ever achieved in her Fortune 100 company. She had been at her organization for eight years when she was told she was in consideration for a regional executive role. She was beyond excited and geared up for the almost eight-month selection process, where she was being evaluated against two other candidates. Though she was exhausted from the politics and the pressures of her previous eight years, she told herself if she could get through the next year, it would all be worth it.

At the same time, she was ignoring the small pains in her hands and the growing headaches. She took more pain relievers and kept going. Then, five months into the audition process, she was at a company retreat in Brazil and she woke up one morning to find she literally could not move the right half of her body. She lay in bed for almost an hour before she was able to inch her way to her phone to call for help.

A few weeks later, she found out she had a heart condition. She decided to keep her condition to herself, but shared with her leaders that she needed to reduce her travel for the time being to address a personal issue at home. When I met her, she was two weeks away from hearing about the role. I asked her why she wanted it so badly, and she said she wanted to be the "first." But when I asked her why that mattered, she realized she didn't know. She started asking me if I believed the politics were different for women of color, and if the men she was up against felt as alone as she did. As we talked, she admitted she was on guard almost all the time and her need to perform may have contributed to her ill health.

Blanca got the promotion and, in the year since I first met her, has changed a lot of her lifestyle. She has set better limits and relocated near her parents, so she has support when she does have to travel. She also journals every day and takes walks. Getting ill, she told me, was a wake-up call to what was important to her.

RACISM'S DIRECT IMPACT

A few women I met are pausing and acknowledging the weight of the system and listening to what their bodies are saying. Dr. Rachel Rosenfeld, a California-based psychologist, says denial and the trauma of racism sublimates and becomes stored in the body. That toxic energy must go somewhere, and eventually the energy weakens our immune system.

Experiencing a racist incident is similar to feeling too much information too fast, like a car accident. Your body can't process the shock and trauma of what is happening, so it blacks out or shuts down. People are prone to either fight, flight, or freeze in these situations. Most freeze. WOC feel unsafe and don't have support to talk about what is happening to us, so we get sick.

In therapy, Dr. Rosenfeld helps clients through a series of memories in a managed fashion. Clients can experience a whole range of feelings in an "organized, titrated way." Simply put, she helps clients complete the integration process to release the emotion and energy. In her view, WOC are sick because the nervous system does so much work to manage the stress. WOC need ways to process the racism they face, and they need social and emotional support when they are made "other." In therapy, they don't have to go through it on their own.

Dr. Rosenfeld also shared that more WOC have started to see her after the racial protests of 2020 to talk about race and trauma, feeling that there is something wrong with them. As the racism is acknowledged by others, these women shift to various stages of grieving including denial, anger, bargaining, depression, acceptance. Before this, they spent years wondering if they were to blame, but they are finally seeing and feeling that it was not them at all. The process is slightly different by racial and ethnic groups, but she is aware of patterns. With South Asian women, she sees more feelings around sadness. The Black women who come to see her often want to express their anger.

One way people of color manage through racism is to develop codependency. They feel like they shouldn't have any needs or wants. Dr. Rosenfeld uses methods from Pia Mellody, an internationally renowned authority on codependency, who focuses on shame reduction to allow people to release the shame they carry energetically. Dr. Rosenfeld says that POC have absorbed other people's shame for generations. We can't metabolize other people's shame because it doesn't belong to us, so we need to do active work to release it from our bodies and shed it, as with the personal delusions in Chapter 2.

REDEFINING SUCCESS FOR HEALTH

Ciel Grove, one of the shamanic healers I have come to know and work with, suggests that chronic illness happens when women are out of alignment. She believes many high-powered, successful women have too much yang, or "doing" energy. We fight so hard to get to these jobs, and then we fight to rise. We put in effort and sacrifice who we are and what else we might want in life. Many of us don't have time to stop and question whether we are on the right life path. When we finally take a moment to pause the flow of constant motion, our inner voice can finally surface, and we start to see success differently. Our new definitions of true success turn out to be much more nuanced than just chasing new heights.

It reminds me of "svasthya," the Sanskrit word for health. Svasthya defines health as "coming into one's own self or realizing one's own self." You are in true health when you are rooted, comfortable, relaxed, and prospering. That word and definition have sat with me for a long time because that is how power can be defined for WOC.

Jeanny Chai is the embodiment of this. As the founder of BambooMyth.com, as well as a coach and speaker, Jeanny has a mission to uplift and empower Asian professionals to rise in leadership. She spent years in the tech industry, working first at a large company and then as an entrepreneur. At thirty-nine, she found herself at her lowest point, newly divorced, a single mom, a new tech entrepreneur, and feeling like she was ten years behind the rest for her Stanford graduating class. She was pushing herself beyond her limits and questioning what she had to show for her career. Then Jeanny found out she had three growing malignant tumors. As she worked with her doctors on a treatment plan, they were very honest, telling her that medicine had come far, but she needed to manage her stress or the tumors and cancer would come back. She says those conversations with her doctors were life-changing.

As she fought through recovery, she realized her definition of success had been tied to material wealth and what her Asian parents had taught her: to obsess about being good at her job and being a good wife. In Corporate America, she had adopted her company's version of success, sacrificing her needs and wants to serve clients and making sales targets at all costs to show results. She felt constantly overworked.

The cancer helped her see she needed to reevaluate the idea of pushing so hard. The corporate and family delusions she'd absorbed, and the pain and shame she'd endured, had made her sick. After her doctors told her she had to manage her stress or relapse, she sat down and wrote her bucket list. "I gave myself stuff back," she says. "Stuff I wanted to do as a kid. I got a pet, and I took singing lessons. I started liking myself." She became intentional, asking herself if she was enjoying each and every thing she did. She realized she had options. Cancer was her motivation to live *her* life.

Jeanny stopped pressuring herself. Her definition of "achievement" changed. She stopped focusing on measuring up and getting to the next rung, and instead concentrated on getting healthy. Her inner voice came back and helped her see she had to let go of the loud voices of her parents. "I wasn't being 'me,'" she says. "I did what I thought my parents and my culture wanted and what Corporate America and society wanted, but I was miserable."

The gift that emerged for Jeanny was similar to mine. We both paused, listened to our bodies, and changed our lives. The voice of her body became so loud that she had to listen to what she genuinely wanted, and shed what she'd been told to want. Now her definition of success is based on inner truth, alignment, and authenticity. Cancer helped her see that achievement is watching her kids grow up. The most beautiful part I have taken from her story is that purpose and contentment come from within.

HOW TO LISTEN TO YOUR BODY

As women of color, we need to listen to what our bodies are saying, even if we haven't decommissioned the programming and delusions in our heads. When our minds will not listen, we can trust our bodies to speak with a full voice, albeit a different kind of voice, and deliver the message we need to hear.

It is believed peacocks have an almost magical capability. They can digest poisonous plants and snakes, things no other animals in their surroundings can. In fact, it is believed that digesting these toxins may contribute to what makes peacocks' plumage so beautiful. They may have adapted this capability over centuries. As women of color, we need to find ways to process and release the poisons that we've had to swallow, including microaggressions, racism, and fitting in. We must know that, in having to maneuver around them, we can often develop our own magic.

LET YOUR BODY TELL YOU.

The bestselling novelist Tom Robbins says there are two mantras in life: yum and yuck. I recommend thinking of decisions along those lines. In our society we listen to facts and data, but we don't give enough credence to the other types of knowing. We aren't taught to bring our gut, body, or intuition to our work lives. When you are expected to make a decision, take on a task, or meet with someone, ask yourself, *Does this make me feel yum or yuck?* He suggests your body will usually let you know.

In other words, does thinking of taking an action make your stomach hurt? Do you get flushed or start to sweat? Do you clench your teeth or tense your shoulders? Do you feel stressed? Start to recognize the small things your body is doing to tell you if a situation, interaction, or decision is bad for you.

GO BACK TO WHAT MAKES YOU HAPPY.

Dr. Valerie says that for us to heal trauma and shift the game, we need to go from surviving to thriving, and from "How much can I bear?" to "How good can it get?" We do this to move away from delusions around success and achievement. In order to do this, we need to remember who we were prior to getting to work, so we can reclaim our authentic desires.

Sit down and make a list of the things you enjoyed doing as a little girl. Do you do any of those things as an adult? If you could pick one, what would it be? Try doing it. Jeanny got a pet. I started cooking more because I loved it, even as a child. It doesn't have to be a big, dramatic trip or undertaking, but by going back to the things that made you happy when you were younger, you can start remembering what you liked to do and carrying it into your adult life. You can then decide how you might add to or change that list as an adult.

FIND YOUR COMMUNITY.

Community is where we draw our power. Being witnessed is a crucial step to releasing trauma. By seeing the same patterns and shared experiences in other WOC, we will start to release the false beliefs that the issues lie with us. You'll read a lot more about how to do this and why it is so important in Chapter 7.

FIND WHAT WORKS FOR YOUR BODY.

Find practitioners who can help you manage your stress and heal your body. For some this may mean doing meditation or yoga, or working out at the gym. For others it may be finding someone like Queen Afua to help heal the body from the inside.

To address the trauma, most practitioners I met said we need to let our body feel the feelings so that we can release them. Ideally, you should find someone skilled in somatic work to help you. In addition, breath work can sometimes help us to move trauma and develop gratitude and kindness toward ourselves. Doing a simple meditation practice each day will calm and focus the mind, and still the thoughts.

If all of that feels too unfamiliar, do two simple things. First, drink more water. Almost every practitioner I have ever worked with suggests that we don't drink enough water, and water is important to feel flow and to not have energy and toxins get stuck in the body. The other simple thing is to get outside and move. Just being in nature and walking can help shift us out of unhealthy feelings and reset our minds, energy, and nervous systems.

••

I still think back to the questions Dr. Heather asked me about what made me happy, and would I be okay doing nothing, and could I see value in just being? At the time, I was overwhelmed and ashamed, and felt in some ways like she had not only seen me but seen through me. Her questions may be terrifying to some like they were for me, but once we start to accept that we know the answers, we get to leave behind everything that is not needed in our lives and strip away all that is keeping us in a conditioned mindset. We get to decide what matters. And that is easier than doing what we are told.

The relationship and communication between the body and the mind are rebuilt slowly. You build trust as you go. Ciel taught me that the feelings that emerge from the body are written in a language we don't remember. We come into the world speaking this language, but we forget it because of our conditioning.

I took this approach during my treatment, and it helped me

reframe my situation. As we were trying to get nearly toxic levels of lead out of my system, a common issue that accompanies Lyme disease, I did weekly chelation sessions. The doctor would inject me with chemicals that pulled out the heavy metals and lead that had built up in my bones since childhood. The procedures made me very tired, and, in an effort to reframe these procedures, I started to do a ritual before and during the chelation. I imagined all the things my body was detoxing and letting go. As I tried to do that, I started to think about the ways that I wanted to grow and change.

In order to redefine power and create the success we want, we must listen to our physical bodies, as well as our inner wisdom. The new paradigm for the ways we want to work, and even the structures and institutions we want for the future, is found within. And similar to the mechanics of my chelation protocol, the vision is inside us; we only need to pull it out.

CHAPTER 5

THE JOB WITHIN THE JOB

When I was in Oakland, I met Suraya, a young millennial Afro-Latina. She blew me away when she shared that, after less than five years working for highly desirable name-brand agencies, she had started her own firm. She wanted to hire more WOC and have more influence on the topics and clients she worked with on issues of social impact. The business had grown in leaps and bounds, and Suraya was at a point where she was trying to manage the demand while still delivering the quality she had come to be known for. At the same time, she had so much gratitude for being able to be on her own and learn her own lessons. Even with all the obstacles, she told me it was freeing to not have to conform to those around her, and to be able to decide which projects she wanted to work on.

Suraya explained that, after graduating from college, she'd been hired by one of the largest and most coveted advertising companies in the world. Even though she was the most junior member of her design team, she would often weigh in with clients about how young women or communities of color might respond to a product. Since she was the only diverse voice in the room, her opinions carried weight early and often. Suraya was promoted quickly and became the designated person in her large office for issues of belonging and inclusion. She didn't ask for the role, and sometimes even tried to give it back, but there was no one else to take it.

A few years into Suraya's tenure, her company went through a merger. The resulting leadership team was made up of white male thirtysomethings. In a town hall to meet the new team, Suraya asked how these leaders planned to bring a full range of voices to customers if the group itself wasn't diverse. No one on the panel really answered her question, and the room mostly went silent after she spoke up. After the meeting, many people came up to Suraya and thanked her for being brave, and their praise translated to a sense of dread for her. She says she quickly went from "pet to threat."

She knew she'd asked a hard question, but had she crossed the "line"? Had she said more than she "should" have? Had she stepped into something she'd have to "pay for" later? Confusion set in. She began to doubt herself and became unclear about what her role was.

As the office's inclusion leader, Suraya felt it was her place to ask questions and provide representation. And if she didn't, who would? But after that leadership session, she never again felt like she belonged in her company or had the stellar reputation that she'd had before. Instead, in the months to come, Suraya was seen as a "firecracker and an agitator."

After the incident, Suraya stayed for another sixteen months, but she knew her days were numbered. In hindsight, she wishes she'd left sooner and that she hadn't let her confidence wane. Her doubt and fear made her stop speaking up. When her team members wanted advice about inclusion-related topics, she didn't feel comfortable giving feedback. She started getting recurring headaches and panic attacks.

As I listened to Suraya, I tried to reassure her that she had done all she could. She responded by saying one of the most profound things I've heard in collecting research for this project: "I knew I had to leave. I knew it in my bones, but sometimes, our very presence is protest—our staying within a structure not built for us can be a form of protest." By just asking a question, Suraya felt she had made everyone uncomfortable. She had put on the table the very thing they knew they had to fix. "In the moment, I may not have known that, but now I do," she says. "Even if I was sidelined, I was there, and they knew I saw through them."

I was impacted by Suraya's idea that staying is protest, because it reminded me that, just by being in the system, women of color become lightning rods. Our very presence and our rise are a refusal to accept the status quo. But, by taking this point of view, this also means we are most likely dealing with a lot of stress and fight-or-flight chemistry in our bodies. Suraya's story really brought home for me the heavy toll being a WOC in corporate spaces can place on our bodies and our mental health. As we talked about in Chapter 4, this is what causes many WOC to get sick. Being inside these systems that don't reflect us is monumental work in itself.

This chapter is about understanding all the work we take on as WOC outside the roles we are hired to do. In addition to the feelings of incongruence, of not being seen or heard in the system, there are a lot of extra tasks WOC are asked to do because they're the first, few, or only. Not only are we executing the role we were hired to do, but also we often find ourselves the sole voice on current national events, and sometimes race, in our workplaces. Because of this, many of the women I spoke with use some version of both "visible" and "invisible" to describe how they feel at work. That duality is confusing, and it weighs on us. We feel invisible because we have to hide ourselves. At the same time, we are visible as women of color, so we are often being asked to do extra things. "People see what we are before they see *who* we are," one executive said.

We have the burden of taking on the extra responsibility that comes with success and high achievement, and most women just accept that these requests come with the job. But it's time we push back and help companies see how much extra work we do. Some of this work is small and infrequent, like speaking to new recruits or being asked to talk about our career paths. Tasks related to race are more complicated. Because race is such a challenging topic, we have to navigate intense politics and penalties in these roles. In order for us to have more agency and power in our work as WOC, we need to understand the full extent of how much more terrain we have to traverse in our workday than our white peers.

UNPAID BURDENS

Most of the extra tasks we are expected to do at work are unpaid and not something our peers are ever asked to do. The women I met feel like these tasks come with the job and are the price of admission, but they shouldn't be. There are many housekeeping and "pink chair roles" like culture-building activities that WOC are told to take on. Pink chair roles are non-revenue-generating responsibilities outside of HR. Women who take on these tasks find themselves working what amounts to an unpaid "second shift." Jennifer Kim, a start-up advisor who works with leaders on people operations, notes that it's the kind of caretaking women naturally assume—picking up the "maintenance" tasks, paying attention to people's emotional needs—that is the work most men see as "soft." Our names come up when these roles are being discussed, and when the company wants help. Most companies expect us to play these roles, but they don't compensate us for them by helping us move up the ladder.

One of the biggest unpaid burdens and invisible requests women of color face is code-switching. Because white people own the dominant culture in corporations, most women of color spend significant time and energy adapting themselves to fit into these spaces. But we don't always talk about the toll of code-switching and how often WOC must adapt to fit into white dominant culture in order to survive at work.

We code-switch not only in how we talk and what we talk about, but also in how we dress, how we stand, how we shake hands, how we network, and how we present ourselves. We code-switch in ways we sometimes don't even recognize anymore because it has become second nature to try to adapt. One woman who worked at a large bank said it well: "It only hit me how much I code-switched when I was supposed to attend a networking event one Wednesday night, and my body just didn't want to move. After spending twelve hours doing it all day, it really hit me how I could not code-switch for even a few more hours, even if there was food and alcohol involved."

I hosted a dinner with some senior white male leaders and realized they didn't understand the extent to which WOC code-switch in corporate spaces. I asked them to close their eyes and think back to being overseas, where they'd have to conduct all their meetings and dinners in another language for maybe ten to twelve hours per day. Most of the leaders agreed those trips were hard and exhausting. By the time they got back to their rooms, they were wiped out. The stakes are high if you say the wrong thing, the stress is high, and the experience is emotionally and physically draining. Some of the men said they typically need a few days to recover after a trip like that.

Most of you reading will know what this feels like. But, until we can help our colleagues and white leaders understand the extent to which this happens and how it shows up in our daily interactions, it can be hard to have them walk in our shoes. For now, it is important we recognize that having to adjust our behavior and our language on an hourly basis takes a toll similar to translating a second language in our heads. If we are going to be happy and healthy in the long run, sometimes we just need to be true to ourselves and how we show up in the world. When the toll feels too high, we need to listen to how tired we are and give ourselves permission to rest and opt out of certain expectations.

To have agency over the work we're handed, we have to ask more questions and point out how burdensome these tasks can be. Then we can shift to being more fully in our power. We have to get used to truly listening to our minds and bodies in order to discern when it feels like something we want to do vs. feels like a burden. Sometimes we can't tell before we say yes if it is asking too much of us, but we feel heated or resentful after we do the work. Paying attention to that resentment can help us figure out when to say no in the future. Although we can't say no to everything, we should point out when the work is an extra burden. The more we say no or ask that roles get shared with other members of our teams, the more the invisible work we do will be visible to those around us.

At work, the job within the job is very often imposed on us. We are expected to be the voice of inclusion, and we are expected to code-switch, but what about the roles we want to take on because we are change makers? As we talked about in Chapter 3, WOC have a desire to give back. We greatly respect the legacies we emerged from, the path makers who came before, and we want to continue the tradition of making a difference in the world. This usually shows up by us taking on one or more of the five archetypes below. We usually take on more if there are no other WOC to share the tasks. The impact of these extra roles can be huge, depending on the individual and the situation. There is nothing good or bad about finding yourself in one of these categories over another, but all this extra work creates a job within the job. Identifying the role you tend to play, knowing you have a choice, and learning how to navigate within your archetype can help you to manage your workload, loneliness, and exhaustion, and give you a greater sense of agency over your work life.

ARCHETYPE 1: THE REPRESENTATIVE

DEFINITION: You represent an entire race or culture within your organization due to being a "first" or "only."

Roxy, a Haitian immigrant working for a giant consumer products company, explained to me that she was often the first Black person some of her colleagues had ever met. She smiled as she said it, but I could also see it caused her great angst. After a few first encounters, Roxy realized, "I'm shaping someone's perception of what a Black woman is. They have literally never met a person of color before. That's a huge, unexpected burden I must carry."

The emotion Roxy unpacked as we spoke is that she felt pressure to be a perfect ambassador for all Black people, and she often finds it overwhelming. How many of us are in the same boat?

Some of the other women said this carried over to things like how they dressed and even what foods they brought to the office.

One Black woman shared that, even though she loves watermelon, she will never eat it in the office or at a company picnic because she doesn't want to propagate stereotypes. She not only felt the responsibility for upholding her behavior to a higher standard, but she also felt the need to confront stereotypes that were outside of her control. In a paper presented in the *Cornell Law Review* titled "Working Identity," authors Devon W. Carbado and Mitu Gulati note that minority professionals tread cautiously to avoid upsetting the majority group's sensibilities. Put simply, their research suggests we try to avoid stereotypes and do extra work to prove them wrong. We may not even know we are doing this.

ADVICE: This role is all-encompassing, and WOC always feel the weight of these intense levels of responsibility and high levels of perfectionism. Remember you are human. You can take a break. You are doing the best you can and should seek support when you need it. The setup at work wasn't created by you, and you can't be expected to fix it all. It reminds me of one of my favorite mantras: "This is about them, not about me." I have learned to say this to myself a lot over the years as I navigate relationships.

ARCHETYPE 2: THE BALLOON POPPER

DEFINITION: Fixer, faces things head-on, a risk-taker and culture changer.

There was a long-standing joke on my teams: even without a nail, I would go looking for something to fix or set right. If there was something that needed to be said, I often played the role of "balloon popper." If I thought it would reduce tension or get our team to the other side of the disagreement, I would ask the question that no one else wanted to ask or say what everyone was thinking but didn't want to say out loud. I wouldn't wait for the balloon to keep filling up before it exploded on its own—I would just go over and pop it.

This role can be greatly rewarding because you are often seen as strong and maybe even the hero when there is disagreement or competing points of view. But it is a very lonely job, and one that often doesn't gain you much praise from or many friends among your white managers and colleagues. Most people would rather play the bystander and let things just happen, but someone who is a "balloon popper" can't let it go. They have consciously or unconsciously taken on the role of being the one to change culture in their companies. They fight systemic racism and all that comes with it.

Nia, who works at a film company, says she is the loudest voice on inclusion by default, even though that is not in her job description and she isn't paid for it. But after finding herself the only Black person in key meetings, she felt a responsibility to speak up. She shared that when a biopic opportunity came up about an iconic Black man, two white men were asked to take the lead, and she struggled with whether she had to "ring the bell" and ask, "Should they be on this project? Is this who should be telling this story?" Even though it is not her role, she feels the need to remind people that it matters who gives voice to stories about people of color.

"White men don't understand that," she says. "They don't have that burden. And there can be backlash." There can be additional opportunities that come with the visibility of taking on this role, but the topic and sometimes the role itself have sharp edges. It's a real challenge for women of color.

A lighter, less intense version of balloon popping may be seen as "truth telling," when a WOC brings issues into the light. In this approach, there doesn't have to be tension or drama when a WOC feels the need to set the record straight or share the facts at hand. What Suraya, the lightning rod and truth teller I introduced in the beginning of this chapter, didn't understand was that, by simply asking the questions she felt needed asking, she was stepping into this role, and it is one of the more challenging roles women of color play at work. We are often placed in positions where we must bring in perspectives about how the outside world will be impacted by certain situations, statements, or products created by our companies.

I found myself being a truth teller often in my career not because I wanted to, but because my expertise was often questioned by my clients due to my age, height, ethnicity, and background. This took the form of new clients wondering how old I was, and if I had the requisite expertise to be advising their companies. I remember meeting a new client and taking two steps into his office only to have him say, "If I had a daughter, she would be older than you. What can you possibly have to tell me?"

I understand why, for him, I didn't command authority. I was a new partner and in my early thirties. I looked young and small in stature. But I had spent a few weeks with his staff and learned that they were afraid to bring him problems or raise concerns because of his temper. Some of the issues had quick fixes if someone was brave enough to share.

I remember swallowing what felt like a big knot in my throat and saying, "If in twenty minutes I don't tell you something you don't already know about your operations, I will give you the rest of your hour back and never ask you for another meeting." He spent over two hours with me that day, and he became a big advocate of mine. This happened because I used the transparency and candor of truth telling to deliver my messages. It became a defining quality in my leadership style.

ADVICE: How you deliver your messages of truth may matter more than the truth you speak. Truths can be extremely hard for people to hear. Finding the right recipe to tell someone their baby is ugly, but in a palatable way, is the goal.

Bring in analogies your white colleagues will understand and speak to the impact on the people around them. Use emotive language but speak with clarity, stay calm, and be equipped with the facts. In her book *The Person You Mean to Be: How Good People Fight Bias*, social psychologist Dolly Chugh talks about this as "heat and light." She says when you speak to people or try to create change using light, you are using approaches that make them comfortable, influence them, and educate them. When you take a stronger, heat-based approach to force change, you worry less about how people hearing the message feel. There may be times when we need to use one over the other, but both are available to us.

The balloon popper is by its nature confrontational, and therefore it is one of the loneliest and hardest to play. If you choose to take on this role, be prepared that people may react before they listen to you. This can be easier if you have senior champions who will support you, even if they do it quietly and discreetly behind closed doors. Make sure you have the strong networks we will talk about in Chapter 7. You will need safe spaces to share your incidents and seek support.

ARCHETYPE 3: THE SEED SOWER

DEFINITION: Your focus is solely to be the "first" to get a seat at the table.

Some women don't want to truth-tell or "pop balloons" all the time. Some want to be "seed sowers" and just get to the table. Over lunch, one woman told me, "My role is not to turn tables or upset the status quo. I am just getting to the table. And I am going to put my feet up once I get there and rest. That is my work. The women who come after me, they will make space at the table or change the table. I am just planting seeds."

I literally dropped my fork when she said that to me. It had never occurred to me that some women of color felt that way. I had always assumed all of us were trying to get to the table so we could make change and make space for others. But she explained that the fight just to reach that point was exhausting, and she had nothing left to give. I spent days thinking about this. Maybe she saw herself leading by example and serving as the role model for those around her. In some ways, what she was saying is similar to what flight attendants tell us. You need to put your mask on first and then maybe you can help others.

I now realize that not every woman has the emotional resources to push on the system. The women who are tired yet still push to get to the table, believing their example alone is making the change, also need to be recognized.

Many of us may want these women to do more, but some of them are already doing more than enough by supporting extended families, doing work outside their jobs in their communities, and even taking on local opportunities like school board roles and organizing resources for local campaigns. For some women, being the first, and mustering up the courage to get to the table, *is* their way of giving back.

ADVICE: This is an interesting role. As more companies and employees talk about structural racism, there may be less tolerance for WOC as seed sowers. As a society, we are expecting WOC to show up differently at work, and to do more for the advancement of all WOC.

The choice is still yours, but know there may be new backlash for not making it part of your work to bring others to the table. This isn't necessarily fair, but things change as our conversations evolve. It takes so much work to get to the top, so some women may need a minute to sit down and relax their feet, but in today's climate the break may be short.

ARCHETYPE 4: THE SAGE

DEFINITION: The unofficial WOC to whom employees and sometimes human resources go to for advice on advancement, race, and diversity-related matters.

My friend Stevie was the first Black female partner at her law firm. To make sure others didn't have to struggle like she had, she organized all the women of color in her company by city and asked the most senior women in each geographic region to host in-person meetings so they could discuss the topics on everyone's minds: inclusion roles, unconscious bias, wellness advice, and the current political climate. She wanted to help less-tenured WOC learn the unwritten rules and understand the politics of her company. She realized many of these WOC didn't have places to go to learn how to navigate issues like politics.

Many of the women of color I met were one of only a few senior women of color in their companies. They often found themselves as "the sage" for many other people of color, and the voice of reason on current events and political affairs. They became the beacon, the support for others around them, and the de facto guide for their department or group—and they were also expected to teach other WOC the ropes and help them navigate corporate culture.

One woman shared that, as a new engineer, she publicly disagreed with a senior leader who didn't like to be questioned. The only other woman of color took her aside and shared that she was sorry for not explaining the written and unwritten rules of the company's office culture earlier. She felt a great responsibility that this was her failure by proxy. Another WOC talked about how she was expected to speak with other WOC about appropriate dress code, even if this should have been done by HR.

The sage role can be a less personal choice than the other roles. Some WOC voluntarily take it up, while others have it thrust upon them. No one pays women to take on this role, and for some, the burden of this extra weight becomes significant and heavy. Research supports that many women in first, few, and only positions can get overwhelmed with having to advise the number of junior colleagues looking for WOC mentors. Sometimes, people gravitate to you, but you aren't taking definitive action or making statements to cause this. If you find yourself in this position, you may not have as much control. The amount of time you spend in the sage role can determine if it is fulfilling. It's easy to get burned out. In some organizations, where there are few WOC at senior levels, this can turn into an almost full-time role.

Many of the Black women I met talked about having to take on extra work educating white leaders and staff after every national story about excessive police force against (and often murder of) Black individuals. The sage often feels moved to educate, and yet this is further complicated because research suggests Black professionals have "to be very careful to show feelings of conviviality and pleasantness, even—especially—in response to racial issues." The study goes on to explain that emotions of anger, frustration, and annoyance were discouraged even after these heartbreaking events.

And although maybe this pressure has slightly decreased since the study was done in 2015, some of the women I spoke with are still struggling with how much to fit in versus stand out.

I will never forget a Black human resources leader sharing that she found it hard to manage her own emotions of anger and sadness as a Black mother. She was often forced to find the words and energy to educate her colleagues about what was happening on the television was not outside of her reality, even though most of them felt it was not something that would ever happen in their town or city. Many women talked about the weight of having to educate others while also finding ways to keep their emotions at bay.

Other sages explained how they had to stuff their anger after the 2016 election and at the same time hold space for many white women who were visibly upset by Hillary Clinton's defeat. They found it hard to find places where they could process their own grief and do their assigned jobs, let alone engage with others to process their own dismay and disbelief.

ADVICE: Some women who play this role feel it is part of their contribution and how they give back to others. They do the sage role willingly and often with great pride. Other women who have this role thrust upon them need to know their limits. Helping others usually comes naturally to people in this role, but remember to set boundaries and take care of yourself. You can help others only if you don't let the weight of their needs take over your self-care. These women need to find others to share the burden. There is power in finding a network of sages and having group discussions rather than one-on-one interactions with every single new WOC. Use technology and share resources like Stevie did to get greater reach with less effort.

ARCHETYPE 5: THE INCLUSION LEADER

DEFINITION: You have a formal role to represent inclusion matters for the company or department. I wanted to add this role here because, although you may be compensated, many women in these roles are usually doing work beyond what they were hired to do.

The rise of the inclusion leader in formally paid roles since 2019 has meant many more women of color are being called to senior roles in order to define culture. This is a great thing because it is our voice that is needed to understand the issues and make change. It is visible and paid work. It's a struggle, though, and many women in these roles feel like companies want to do the right thing but don't have the structures, processes, or budget to support it. Many may have the title and salary but not the staff or budget, so to reach success they must do a lot of extra, unpaid work.

Despite the dismal experience many women of color have when they take on inclusion roles, the women I spoke with almost always wanted to help their companies to improve on this essential facet of their culture. They felt a responsibility to do the work and assumed it would be fulfilling. But almost all of them wanted to choose the roles and tasks they took on and expected to be recognized and rewarded for tackling what can often be difficult, thankless work.

ADVICE: This role needs to be set up in the right way. If you hold a formal inclusion role, make sure you ask for the ear of the CEO and the executive team. It's important to have staff and a solid budget. Ask questions, and make sure inclusion will be part of the company's core business, not merely an afterthought.

Without the right team, support, and influence, your workload may be unmanageable and have limited results. I know women who have recently accepted these roles only to be sadly disillusioned a few months in because they do not have the support to make the changes that need to be made. These women have high levels of burnout, and they feel strongly that what was promised to them when they accepted the role is not what the company provided them once they were in the seat.

THE FINAL CHOICE IS YOURS

These five archetypes cover most of the ways I've heard WOC take on work outside their formal job description. The upside is, whether you are taking on one of these roles, multiple roles, or none of them because it's not your work to do, the list below can help.

If you do choose to play these roles:

1. **NEGOTIATE.** Try to negotiate when you say yes. Ask for what you want and make trades if necessary. Do it with poise and tact, but try to be clear about what you are willing to do and what feels like too much.
2. **KEEP TRACK.** Try to manage how many additional roles and tasks you take on. One only has so many hours in a week and, if some of the additional roles don't get credit or aren't seen as valuable contributions, you will need to decide how much of them you want to do.
3. **REMIND PEOPLE OF YOUR CONTRIBUTIONS.** Ask whether the tasks you do, the roles you occupy, and the ways you contribute will feed into your year-end reviews or be seen as examples of your leadership skills. Even if they don't count officially, keep a record of them and list them

on year-end documentation or share them via email when the time is right. Continue to remind people of your contributions.

4. **KNOW WHERE TO DRAW YOUR LINE.** Know what work you enjoy and how much energy you have to give. Know when you are being taken advantage of. Situations won't resolve themselves. It may be time to look for brighter horizons.

If you are engaged in these activities remember that even in a world where there is more pressure than ever being placed on WOC to do and say more, you *do* have some choices in deciding whether, and even how, you take on many of these roles. But while they may sometimes feel thankless, you could very well be changing the course of culture.

Colin Kaepernick is the first person who comes to mind when I think of someone who took on a greater responsibility outside of what he was hired to do and, in doing so, made a huge difference. Although he couldn't have foreseen the future and the cost his actions would bring, he kneeled in protest during the national anthem. Kaepernick has not played in the NFL since January 2017, and his career was cut short when no team would sign him. Following a season of protest, he became a polarizing political figure. But this changed the course of his life; he became a force for social justice. Although he may say he didn't have a choice, that it was the right thing to do, he did in fact decide to step into a role—and limelight—that went well beyond just football.

Other sports figures, like Naomi Osaka and Serena Williams, have followed suit, making their voices heard for equal rights. LeBron James articulated he felt the call to do more outside of his role than just play basketball after he was told to "shut up and dribble" by Fox News host Laura Ingraham. He felt the need to speak about race and the challenges of being Black in America. Viola Davis and America Ferrera are also using their platforms to talk about issues related to being WOC in the world, including discussions around representation, immigration, and human trafficking.

It feels like we may be entering a world where many will feel called to do more than their assigned roles and use their power for influence. Our voices and opinions will be just as important as the work we do. In fact, how we use our voices to amplify issues around social justice and inequality may become even more important than the jobs we have.

We may be noisy in getting to or reaching the table, but many of us believe our work within corporate spaces is about disrupting the status quo and making change. Whatever you decide, be confident, understand the added work that will likely come along with the role, continually check in with yourself, and try to enjoy the process.

CHAPTER 6

OVERHAULING A CULTURE OF AGGRESSION AND INACTION

Kara, a senior executive and twenty-year veteran in the financial services industry, shared a simple but quintessential story about microaggressions that has stayed with me. A junior data analyst and fellow woman of color approached Kara after a conference. Kara heard the analyst's voice crack as she began to speak. "I am not sure this is really work-related, but someone said something to me today, and I can't figure out why I'm so upset. I don't know who to ask."

She explained that one of the men at the conference, an older white gentleman, had pulled her aside during a break and asked her a series of questions that bothered her: "Why would *you* ever want to work in finance? Why would a Moroccan woman like you choose this lifestyle and these hours? It would mean being away from your family a lot. Don't your parents expect you to be married soon? People from your culture marry young, right? You must be looking for a husband or a quick way to pad your résumé, because there is no way you belong in this industry." *What?*

The data analyst had beat herself up at the time for not saying something, and she was still doing it three hours later. "I just want to ask what you would have done or if this would have bothered you?" she asked. She was irritated and upset, and as she fought back tears she said, "I don't know why it is bothering me so much."

But Kara knew why and decided she needed to play the "sage." She took the data analyst out for coffee so they could talk. Both of them spent an hour talking about microaggressions, how the topic was absolutely work-related, and how it wouldn't be the first or last time something like that would come up in the data analyst's career. Kara knew this from experience. She shared all the things she could think of that she wished she had known when she first started her job.

Early in Kara's career, a senior executive client had pulled *her* into a conference room after a meeting. As they were sitting around a big table, he asked why she wasn't married and whether her parents had put pressure on her about it. Kara, who is part Filipina, said she was surprised that her client implied anything about her parents, especially since she had never spoken with him about her culture or her upbringing. Kara remembers thinking, *Why does he think this is an appropriate topic to bring up?* She didn't know him well, but since he was a client, just telling him that it wasn't his business didn't feel appropriate.

This type of overreach had happened a lot to Kara, maybe even on a daily basis—but this specific conversation stayed with her because it ended with him asking if he could give her some unsolicited advice. Instead of waiting for her to say yes, he continued, "I think you will have to make choices. You can't have it all.

Women can't be smart *and* pretty, especially women like you. You will have to pick one. If you want to get married, you should pick wisely."

Kara spent weeks wondering why she froze and what she should have said. "For a long time," she said, "I wondered if I had somehow opened the door for that conversation."

You will likely find yourself in situations where you feel your power being sapped. People will say the one thing that triggers you, pushes your buttons, or makes you question your worth. The microaggressions sting and anger us. We may feel the need to address every comment or to overexplain, so much so that we become defensive. Other times, an incident will occur, and we may be stunned to silence, almost paralyzed, voiceless, and unable to respond or stand in our power.

Research nFormation conducted with Fairygodboss in 2021 found that 60 percent of WOC feel their companies are not properly prepared to handle racist incidents in the workplace. The perpetrator usually makes some comments and goes on with their life, while we are left wondering, stewing, and reeling as the data analyst did above. And if we raise the matter with our supervisors or HR at our companies, we are often told we are too sensitive. They say there is not much they can do. It probably wasn't ill-intended or was "just" unconscious bias.

As WOC we are left to deal with the aftermath of microaggressions with little recourse or support within our companies. There should be more consequences for the norms people set, the infractions they commit, the culture they create, and what companies tolerate or even condone. We need greater accountability. This shouldn't be optional; it should be the rule.

It feels like more WOC are beginning to take a stand and call out incidents on social media or within their networks. In the months and years to come, I hope companies will see and name these microaggressions for what they are: a form of racism.

THE IMPACT OF MICROAGGRESSIONS

On the surface, the classic microaggression may appear to be a seemingly insignificant slight, especially from the person unleashing it. But for women of color, there can be lasting effects. As microaggressions accumulate, our working environment begins to feel toxic and our stresses multiply.

It's well-known that negative comments have a stronger impact on us than positive comments. Since we're wired to hold on to criticism longer than compliments, it sticks with us and continues to inflict damage long after the comment has passed. "This is a general tendency for everyone," said Clifford Nass, a professor of communication at Stanford University. "Some people do have a more positive outlook, but almost everyone remembers negative things more strongly and in more detail." He explains that there are physiological as well as psychological reasons for this. The brain handles positive and negative information in different hemispheres, and negative emotions require more thinking and processing time. We tend to ruminate more on unpleasant events and use stronger vocabulary to describe them, so know you are not crazy for harping on them in the hours and days to come.

Microaggressions make us wonder whether the person is intentionally disparaging us, or if the paranoia is getting in our heads, and if we should be second-guessing every comment. It's an exhausting, eroding process. Catalyst, a globally well-known nonprofit focused on advancing women to leadership, says that microaggressions weigh on us like an "emotional tax." This "tax" can be the price many WOC pay for moving through corporate spaces. The research shows that even microaggressions that are so-called jokes result in people of color remaining highly guarded at work. Most of the WOC I spoke with have experienced some form of one of the following three common microaggressions on a regular basis.

TYPE 1: THE "SECRETARY OR BARISTA" MICROAGGRESSION

When the men in the group expect you to take notes during the meeting or to fetch the coffee. One Black woman who worked at a tech start-up said her supervisor always asked her to do the logistics, like booking space and getting snacks for company-wide meetings even though she had a strategy role and nothing to do with event planning.

TYPE 2: THE "YOU ALL LOOK ALIKE" MICROAGGRESSION

One woman in advertising said a white male colleague consistently called her by the name of the only other Asian woman in her office. This happened *four* times in just a few weeks, even though she corrected him every time and had worked with him for years.

TYPE 3: THE "YOU DON'T SOUND LIKE OR LOOK LIKE WHAT I EXPECTED" MICROAGGRESSION

One of the Black financial advisors I spoke with said she has had clients brush her aside because they had expected to meet a white woman. Once she clarifies that she *is* the person they have been working with on the phone, sometimes for years, she often hears, "Oh, I assumed you were white based on your voice."

These three are just the examples of what I heard most often. Many of the women I met shared that microaggressions and even incidents of blatant racism happened on a weekly, if not daily, basis. Most of them tried to push these incidents aside or accept that they were the price of being in a white, male world. But as we have talked about, they take a toll on our health and well-being, and in many ways, dealing with them and processing them becomes yet another task or project we take on at work that is unpaid and underappreciated.

WHAT TO DO IN THE MOMENT

Though microaggressions happen regularly, I understand it's hard to know what to say or do in the moment. Most of us are caught flat-footed, especially earlier in our careers. These comments or actions can happen anywhere, at any time, and from almost anyone. Trust or relative comfort with a coworker is shredded in an instant when they launch a microaggression that leaves us feeling shell-shocked and unsafe.

Again, this goes back my favorite mantra that helps when I find myself in a situation where someone has said something bothersome or offensive: "This is about them, not about me." After repeating that three times, if I still feel angry, I say something. If the anger recedes, I try to move on. It's not perfect, but it is my way of knowing if I want to react in the moment.

I have also learned to pay attention to how my body feels. If I feel hot or get that uncomfortable fluttery feeling in my stomach, the incident is not something I will let go of even hours later, and I should address it in the moment.

Let's revisit the examples from the previous section. Those women all shared what they did to address their antagonist, and I want to share their advice with you so that you have options when microaggressions occur.

EXAMPLE 1: The strategy lead booked the room and ordered the snacks, but made it clear she would not do it again. She actually told her supervisor she was not the "office mami." I am not sure this would work everywhere, but he agreed next time it would be someone else's turn. It became a task that rotated throughout the company.

EXAMPLE 2: The advertising manager finally found herself in the same meeting with her Asian counterpart and her male colleague. She went out of her way to sit next to the other Asian woman so that he would have to acknowledge them by their actual names to differentiate who he was talking to. She said, "He now not only remembers my name, but he also greets me by it."

EXAMPLE 3: The financial advisor shared that if her clients seem to have an issue with her being Black, she has decided to fire them. It took her a while to stand up for herself, and in doing so, she paid a cost—she lost commissions with clients. But she decided, "I'm not going to work with people who disrespect me." She was able to do this because it was her own portfolio, so she was the only decision maker involved. No matter the circumstances, if incidents make us uncomfortable, we all need to be able to draw our boundaries.

One young Korean woman shared, "It's difficult to move past the tactic of letting things slide in order to not come off as a troublemaker." She pushed herself to do it, though, when she found herself in a situation where a white male supervisor was talking about how difficult it was to pronounce Asian and ethnic names, in response to a news anchor's butchered pronunciation of the Atlanta spa shooting victims. She cut him off to say, "We learn how to say European names like Schwarzenegger just fine. Why can't you learn these names?"

Another woman I spoke with shared an additional practical, but brave, technique. She let the first microaggression slide to see if it was an anomaly. The second time the same person unleashed a microaggression on her, she would pull them aside and explain why it was inappropriate. The third time, she'd embarrass the microaggressor in front of a group and let peer pressure change their behavior.

These examples show that WOC do push back. They may not do it the first time, but over time and with a well-crafted plan, they will stand up for themselves. It is okay to say something if you feel safe enough to do so. It is okay to use your voice and point out when something feels wrong to you. That is what Kara told the data analyst.

I tell WOC to think about how hard Vice President Kamala Harris must have practiced saying, "I'm speaking," with just the right tone and facial expression when Vice President Pence tried to speak over her in the 2020 VP debates. She balanced being firm and clear while also not coming across as too aggressive or unlikeable. Unfortunately, we all have to practice our responses like VP Harris. Whatever style you choose, you need to practice how you will respond in racist situations.

We have to find ways to push back and stand our ground in tough situations. In some work cultures we may need to do this while simultaneously tamping down our outrage and emotion. My advice is to think about the microaggressions you have faced and spend a little time rehearsing your responses. While we can't stop microaggressions from happening, we can try to prepare for them. I also believe that the more comfortable and active we are in pushing back, the less they will occur in the workplace. Here are a few examples:

"That's not acceptable, Greg."

"You may not understand what you just said, but it's a problem for me."

"I don't appreciate what you just said."

"Please don't say that to me again."

It's also important to talk about the incidents that happen with your support system, be it the women of color you've bonded with, your therapist, your partner, or your best friend outside of work. Getting some perspective helps ease the sting and can move you forward faster.

I also want to recognize that it may not be feasible to push back at all. Some work cultures will allow for this more than others. When we have to look out for our own mental and physical health and assess our financial stability, we may not also be able to call out an aggressor and shame them so they learn their lesson. It is also okay to ask for help. All the work of culture-fixing can't fall on us. So, ask other WOC and co-conspirators for support when you need it.

And here is my advice to our white co-conspirators who are reading this: don't wait for the woman of color to say something about a microaggression. It shouldn't be up to just us to complain. And don't assume someone else will intervene—the bystander effect is strong. People will watch, thinking someone else clearly knows this is wrong or someone else is better equipped to jump in. The result is that a lot of people silently agree that there is a problem, but there's no actual intervention. This makes you complicit in acts of violence and discrimination. Speak up if you see or hear a microaggression. And the best time is right there in the room, as it is happening. That way, everyone can benefit from this teachable moment.

WHEN THE SYSTEM FAILS US, EVEN AT THE TOP

I usually share these ideas and experiences of microaggression with women of color who are newer to the workplace, like Kara's data analyst. But the obstacles around speaking up against the system and issues related to race don't always lift as we rise. The first, few, and only who I spoke with shared that there was sometimes even less of an ability to push back on incidents and infractions at senior levels because there was the expectation that senior WOC would toe the line.

These senior WOC talk about feeling like unicorns in their roles. They know they should speak up when someone in the company is saying or doing something inappropriate or racist, but they are also afraid of calling attention to their differences. They have spent years and sometimes decades proving themselves. Then an incident happens, and they are instantly aware of how different they are from their senior white colleagues. They may be the only ones who are able to recognize and name it, but they may also be remembering that early childhood message of not "rocking the boat."

When incidents arise, other women I met do exercise the power they have earned as a result of being senior in their organizations, but these WOC find that the system and the processes that claim to protect them don't always show up for them. They learn the hard way that these processes protect the company, and even at the senior level, there is less tolerance and patience for addressing issues.

Malia had been a senior leader for ten years when her company asked her to take responsibility for one of the largest accounts in the company portfolio. Three weeks after Malia took the role, she quickly sold the same client the largest deal of her career. To make sure the deal went smoothly, she was paired with another account executive, Frank, who was white and about ten years her senior.

From the first few days, their interactions were rocky. She didn't know Frank well, and he made it clear she was "lucky" to be working with him. In fact, he told her he had the ability to make or break her career. He shared that he sat on the compensation council and had the power to decide her next step and her pay.

Over the duration of the next year, Frank continued to speak down to Malia and would yell at and berate staff, mainly people of color. Malia went through the typical stages of fear, then panic, then rage, and finally angst. She'd never felt so disrespected, and his behavior seemed incongruent to the company's values. His behavior toward her felt clearly race- and gender-based, and she didn't believe he would have been as disrespectful to a white male colleague. This was confirmed when leadership shared that Frank had a history of "small infractions" with other women in the company.

Malia reached out to many senior white women within and outside of the company, asking for advice. They said that it wasn't sexual harassment and wasn't really that bad. They believed the right thing to do was to put it aside and keep performing. This is another issue many WOC face. There aren't enough WOC before us, so we are forced to take advice on matters like this from senior white women, many of whom don't understand the nuance of race in the workplace.

At six months, Malia hit her breaking point. In one of their meetings, Frank pointed at a spreadsheet and made a comment about "actual yellow people being in the countries highlighted in yellow." He made these remarks in front of others, including other East Asian leaders and staff like herself.

Even though peers told her it was essentially career suicide, Malia told the company to take her off the account and filed an official complaint. After all those years of looking the other way when small things happened, she drew her line and took her power back.

The company's HR team determined that it was a "he said, she said" situation and no one else in the room claimed to have seen anything. They didn't put much effort into reprimanding Frank, but Malia felt the effects. In her performance review, she was marked down, her ratings suffered, and she was told she needed to learn to get along with others.

I heard stories from seven C-suite WOC that were almost identical to this one. In situations where these women faced career-defining moments, they were made to feel unsure, unsupported, and powerless. This type of incident made most of the WOC I spoke with leave their roles and walk away from it all, because the company chose to protect itself instead of facing issues of racism and sexism.

When these women needed support for an issue that so many WOC face—support that would have required their companies to rethink the leeway they gave to white men in leadership—they didn't get it. The company processes don't show up for any of these women. They eventually find themselves at odds with their companies and question their values. Similar to the discussion in Chapter 4, these WOC suffer significant emotional and physical tolls because of the stress of working through hostile situations and wondering how they could have avoided the career blowup and done better staying on track.

When I spoke with Debra Soltis, a well-regarded civil rights/employment attorney in Washington, DC, she shared that Malia's story is not unique. When women of color call her, they are generally stunned. "How did this happen?" they say. Many are high-talent WOC with so much promise. They are fast-tracked, and then, late in their career, they encounter a senior leader, frequently a white male, who reveals a preconceived and demeaning notion or stereotype, which is often fueled "more by ignorance than venom."

Even though the women are well regarded, Debra told me they

often hesitate to formally complain. They have been served well by playing the game for so long, but now the rules have changed for them. Those who choose to complain get pushback for the first time in their career. "Why can't you just let it go?" "Why fight so hard?" "You are overreacting." They are, "sometimes subtly, sometimes not so subtly" taken off their track, punished, sidelined, and muted for saying their truth. "They didn't play the role they were supposed to play," Debra says.

In the end, the women Debra works with often decide to threaten or file claims for discrimination and retaliation, and they wind up with a large sum of money and an empowered sense of self for having fought the good fight. But the professional implications are real and lasting. They ultimately move on, with many exiting the corporate world and finding places where they can voice and live their truth without being penalized.

Standing up to the system is a difficult decision for a senior-level WOC to make. Raising concerns through a company's internal HR system leaves many WOC feeling overwhelmed and discombobulated. Debra counsels her clients to always remember that "human resources is not your friend, but is there to protect the company." Reporting cases of inappropriate behavior, whether it is racism, sexism, or even sexual harassment, leaves many women to face the daunting task of having to present claims that are based on allegations that are undocumented, retaliation that is subtle, and witnesses who are reluctant to speak.

Debra echoes that while the claims of WOC in senior positions can serve as powerful indictments of systemic sexism and racism, there's often less empathy within the corporate world for those who have risen through the ranks and then choose to push on the system.

Another woman with a story similar to Malia's expressed her experiences with even more bite. "Once we get to the top, it's as if there is no real seat when situations like this arise. The men around us have less patience for us because now we want them to change *their* behaviors." She said maybe at one point we were their friends or even a pet. But, she said, "It is clear now we are a threat." She went on to say that just like us, they were taught a delusion—"the delusion that they are always right."

The eight women I spoke to, who went through similar stories of fighting against racism and sexism, urge other women of color to document their stories, use formal channels, and not worry about playing nice or by the "company rules." Companies may try to cross your boundaries, confuse your facts, and discount your truth, but there is power in knowing where your line is.

FIGHTING THE SYSTEM

Leaders need to be willing to look inside their companies and be open-minded that it is improbable that any company is void of bias and racism. It's crucial that companies over-index when the first, the few, or the only experience a microaggression or a racist incident. Fortunately, there are some indications that processes are improving, and in the future there may be more allowance for WOC to come forward.

Kia Roberts was formerly employed at the NFL as the director of investigations to review misconduct allegations involving players and employees. After leaving, she created an end-to-end investigative and legal practice, Triangle Investigations, to examine allegations of misconduct at companies and organizations. Kia is seeing some companies shift from "this one incident happened, let's investigate" to proactively looking at the company culture and broader patterns.

Complaints historically used to fall into two areas:

1. This complaint is unsubstantiated.
2. This complaint is substantiated.

Now there is a third: this complaint might be unsubstantiated, but it is inconsistent with the employee culture we want to have, so we should investigate and fix it.

This third category has come in the wake of the evolving 2020 conversations on race, and a new shift in talking about racism at work. Kia says more companies are looking at data and asking different questions. Executive teams are trying to get in front of potential issues, asking, "Why is no one getting promoted in a department?" and "Why is turnover higher here?" Her clients finally want broader-based data to understand what is happening.

Just like the growing discourse on race has shifted the ground for WOC, it has also changed perspectives for employers. There are more *New York Times* pieces being written, more social media alarms being sounded, and more general awareness that practices of the past can't just keep moving forward. Kia says some companies are even setting up whole departments for race-related misconduct investigations.

Kia thinks that many women of color still don't want to come forward, and there are probably more incidents than we realize. WOC are juggling stereotypes, she says, and because they don't see themselves represented at higher levels, they are worried about career trajectories and are concerned their teammates won't support them if they raise issues. Kia explains that you may not be able to come forward at every company. You need to know your work culture and your HR department, and understand where you are in life and what toll escalation might take.

Often, we don't hear about issues of racism at our own companies because of nondisclosure agreements (NDAs) that do not allow WOC to come forward publicly. Data reveals that over one-third of the US workforce is bound by NDAs.

Julie Roginsky, a cofounder of Lift Our Voices, a nonprofit created to eradicate NDAs, says, "There is history and a protocol of protection and a narrative that silences voices and the truth." Processes are not set up for WOC, as most senior executive teams are "pale, male, and stale." She says most people sign some type of NDA as part of their employment contract and don't even know it, and later down the line, severance is typically tied to more specific NDAs.

Many WOC will sign them again after an incident just so they can get out quickly. Most of the senior WOC who shared their stories with me cannot make formal complaints and asked me not to name them. I had to mask their stories because they have some type of NDA in place that stops them from telling their stories and talking about bias, racism, and sexual harassment.

This stops women from coming forward. It keeps the public from knowing when there are issues within companies. And it also makes it hard for WOC to know if there are widespread issues even at their own companies. In addition, most NDAs also include mandatory arbitration clauses, so even when a case *does* move forward, the company can influence outcomes and maintain secrecy more easily. Julie believes banishing NDAs is the civil rights issue of our time.

It's not all doom and gloom. There are success stories about WOC who have fought the system. Some women have left to occupy new, better roles or have started their own companies and nonprofits.

Dr. Akilah Cadet is the founder of Change Cadet, an antiracism consultancy. But years ago, when she worked for someone else, she faced a toxic work situation where her white male boss told her, "I didn't think you were that smart when I interviewed you, but you *are* smart." When she asked him why he said that, he realized it was offensive and apologized.

She decided she would not let it consume her and she would take her power back. She explained that it was hurtful, and that her boss needed to rebuild her trust. The next week he fired Dr. Cadet without cause, and when she went to HR for help they said there was not much they could do about her boss's behavior. "That's how he is," they told her. "We can't change his behavior." That answer didn't sit well with her, and after trying to solve the issue through internal escalations processes, she filed a lawsuit. She said she wanted to stand her ground and hold her employer accountable. In the end, after going to court, she won her case.

Dr. Cadet's goal was less about a payout or retribution and more to create case law, or precedent, for other women in the Bay Area. She felt strongly that WOC are silenced in the workforce and wanted to do something about it. "When negative things happen," she says, "women of color need to show up for ourselves because spaces don't show up for us." It wasn't an easy process, but she is proud of what she did and feels power from being able to speak about her situation freely.

So, what should you do if you are facing a significant issue at your company?

1. Document everything! Document it even if you don't intend to take legal action—it can only help you with HR.
2. Go to HR, but be wary. Kia explains that not all HR departments are ready to address race and ethnicity issues, but you can and should try to bring attention to the issue.

3. Understand your company's process for reporting bad behavior. All women of color at any level should know who to call when something happens, because it will most likely happen. Even if you find it hard to speak up for yourself, you can be a champion for others.
4. Get expert advice from people like Debra and Kia and look for resources like Lift Our Voices to help you navigate.
5. Be patient with yourself and get support. Whether you go forward and push on your company or decide it's better to let it go, either process will take a toll. There is no right answer—only you know what you can do. You will need help. Make sure you have your resources and your support team at the ready before you decide anything.

The current rules and structures within corporations are antiquated and reflect a white, male leadership structure. They are set up to protect companies and cover up issues instead of creating transparency. America's current racial climate is fueling change, but in the meantime, it is important that you are informed, know your rights, find community, and get advice on what to do if you are navigating an instance of racism at your company. You are not alone. There are resources available. Most of all, we need to support one another so more of us can come forward, feel safe, and find our full power at work.

••

We need to start listening to our bodies and leading with all the tools available to us. How can we possibly be healthy and happy if we are severing parts of our identity and numbing ourselves to make our work lives fit into our whole lives?

There is power in leading in new ways. Power that we aren't supposed to tap into because it is a different kind of power. But for us to begin to have a chance of fixing the world around us, we need to tap into everything available to us, actively decide the change agent roles we want to step into, and agree on how we will band together and fix the broken system that wants to keep us from speaking out about its flaws.

PART III

FORGE OUR POWER

CHAPTER 7

THE POWER OF WE

When I was fifteen, the real estate market tanked, and my father's business went under. We lost almost everything. It was hard to pay for basics like gas, heat, and food. One day, I was standing in line with Acha ("Dad" in Malayalam) at the local county office, filing paperwork for public assistance. Behind the plexiglass divider, a white woman leaned over to her coworker and, in what she must have thought was her "inside voice," she said, "Now we have people showing up for help with names I can't even pronounce." I can still feel the shame from that moment seared into my psyche.

Flash forward to the 2016 presidential election. The time after the results were announced was hard for me, as it was for many women of color. I was especially crushed since I had interned in the Office of the First Lady, for Hillary Clinton, when I was in college. I remember going to sleep on the night of the election, believing Hillary had won, only to wake in what seemed like an alternate universe where she hadn't. In a matter of hours, I felt a deep sense of dread, despair, and fear for what was to come. I began to openly question if I was welcome in this country, and if my feelings of not belonging were about to be triggered again.

Before I started working at Deloitte in 1999, I'd lived in Washington, DC, and worked in politics. I had always assumed I would return to policy and politics after a few years in the private sector, but I never went back. After the 2016 election results, I started to second-guess my corporate path, and bigger questions opened for me on how I was positively contributing to the world. I was again grappling with inquiries about who I was, who I wanted to be, and what my work was in this world.

For once in my life, I slowed down and looked around. Was I living the life that fifteen-year-old saw for herself? Leaving my career felt particularly hard when I remembered the financial challenges my family had faced. Because of that time of financial hardship and the beliefs I had been taught by my immigrant parents, earning a large paycheck was part of my innate sense of stability and security.

A number of tough work crises and my growing health concerns combined with the election threw me in a deep spin about whether I should leave my job. I was successful, but I didn't feel like I was contributing to the greater good. I wondered if hearing how other women of color navigated corporate spaces would help me with my decision, but I didn't have a close group of WOC friends with whom I could consult and talk about my struggles around belonging. While I had worked on me, I also needed to find the "we."

In early 2017, I started gathering WOC in an attempt to find some answers. I contacted women I had met at conferences or professional events, and I wrote to women in my network. These WOC introduced me to others they knew. Over time, one-on-one dinners turned into group dinners, and conversations about whether I should go turned into what I should do once I left. Eventually these turned into larger gatherings or salons, with some even happening over Zoom as a result of COVID-19. In these safe spaces, women of color shared their experiences and challenges.

Those sessions felt like pure magic. They were electric and energized me for days and weeks after they happened. They felt sacred, groundbreaking, and comfortable all at the same time. And Konda Mason, the cofounder and president at Jubilee Justice, which focuses on bringing economic equity and sustainability to BIPOC farmers, helped me put into words why they were so special. I had seen Konda speak on a panel about women and power in 2019, and I was so impacted by her words and her clarity on how systems needed to evolve that I reached out to her for an interview. In our discussion, she shared that I was creating "sangha," a Sanskrit term for community. She said we need spaces like the dinners and gatherings to "relearn and unlearn together." She shared that we need to do deep work like rethinking systems together in community, because it is not work that can be done only on our own.

My gatherings helped me leave my role behind and turned into the foundation for this book and my new company. nFormation's mission is to provide a safe space for women of color to share their stories, find support, and dream of a new future. Rha, my business partner, calls this the work of deciding what we believe as WOC. We are working to define our future and our legacy, asking ourselves the following questions:

What do we believe about ourselves?
What do we believe about the structures around us?
What do we believe is possible if we come together?
What do we believe our legacy should be as WOC?

Our initiatives are focused on helping WOC see how they were indoctrinated and realize what is possible if they give themselves permission to see differently. We do this in community and sisterhood. Our members span across sectors, industries, and even countries. They work in corporate and nonprofit roles, state and local government positions, and academia. We even have some members of the military. Across all these spaces, we are often the first, few, or only. We share events for WOC by city so they can find community, and we have a WOC marketplace so that we can buy products and services from one another and support our growing economy. We are also working with select recruiters and companies to place our members on boards and in senior positions at WOC-friendly companies.

By coming together as women of color, we find community. At nFormation, we invite women into safe forum spaces to encourage one another and share their experiences. We conduct discussions to talk about power and how WOC want to show up in our leadership. We ask what is possible if we band together at work, and we talk about ways to move past the status quo.

WHY "ME" NEEDS "WE"

The internal work of finding power is what I call "the power of me." This work is about questioning the world around us and letting go of the messages we have internalized that are not ours to hold. As we do this work, we start to trust ourselves more and listen to our inner intuition and our bodies.

"The power of me" embodies all the lessons from the previous chapters:

Step 1: Recognizing the corporate delusions around you.
Step 2: Shedding what does not serve you.
Step 3: Rediscovering and carrying what makes you special as a woman of color.
Step 4: Listening to your inner wisdom, your gut, and your body to lead the way.
Step 5: Deciding which roles you are willing to take on at work, if any.
Step 6: Learning how to protect your energy and sanity as you navigate the workplace.

But to profoundly change the system and sustain our personal power, we also need "the power of we." We need to find others to feed our strength, to know we are not alone, and to be witnesses to our struggles so we can be fully seen.

Nathalie Molina Niño, an investor, entrepreneur, and author of *Leapfrog: The New Revolution for Women Entrepreneurs*, believes WOC in corporate spaces have been groomed to suffer alone and to muscle through, but gathering and sharing can break that pattern and end the silence and alienation. She talks about the power we can have if we build alliances and leverage our power in numbers. "I don't want us to suffer alone but see other stories and gather so we can widen our vision," she says. "It's easy to silence one of us, but it's impossible to silence all of us."

It is hard to live our power if we don't have our sisters with us to prop us up and be there with support and compassion. We also need to come together because we cannot change power structures on our own—we need to do it together through the power of "we."

Rha often says that 2020 was about tearing down structures and old ways of working, and helping us uncover many of the delusions I listed in the beginning of this book. But she also says we can't just go from tearing down to erecting new ones—we need time to contemplate and to sift through what we want to hold on to and what we want to release. I also think we need time to do our inner work and then to build community among ourselves as women of color so that we can decide what is next.

KILL THE QUEEN BEES

Noni Allwood, who we met in Chapter 3, asked me if I believed there is an additional delusion, that women of color think they must climb the ladder of Corporate America on their own. Do women of color believe they need to be "tough" and "go it alone"? She wanted to know why it is so hard for women of color to come together and be a force united at work.

I heard consistent stories from WOC at all stages of their careers saying that we need to do more to support each other in corporate spaces. A surprising pattern I found in my research was that many WOC actually felt sabotaged by other women at work. When I first heard this, I assumed it might be *white* women not helping WOC, or that this issue actually might not be present for less-tenured WOC. (I do think younger WOC are trying to do this differently and to help one another as they climb. In Chapter 8, you will hear how some of them are banding together to share pay data and help one another with advancement negotiations.)

But overall, when I ended my interviews and asked women if there was anything else they wanted to share, dozens of them would drop their voices and say things like "Can you write about how we get in each other's way?" or "We are helping the white boys win."

When I asked one Black midlevel woman in health care if she had any last comments, she ended our call by saying, "I hope you will talk about how we don't help each other as women of color. It is never talked about, and if we don't talk about it, we can't change it." She was not alone in her sentiments. Another Black woman in finance shared that a senior Black woman had taken her aside in her first days on the job and told her that there was room for only one of them at the company. One of the younger WOC I met shared that, although the WOC ahead of her were helpful to her now, she was afraid that might not last as she rose. She had watched them not support one another at more senior levels.

Even in early 2021, as I was completing final edits on this book, I had a big debate with a group of WOC about whether we trust and lean on one another. A few of the women felt there were places where WOC helped each other, like Black or Latinx sororities, but we agreed it's harder to find stories of WOC supporting one another at work, especially at the midcareer level and higher.

I don't think the reasons behind this are sinister, but I do believe the lack of support women of color give one another is related to the delusions happening in corporate settings. Many WOC are heads down on their tasks and overworked, so they don't have time to help one another. This is combined with an erroneous belief that there are limited seats at the table for WOC, so they need to compete against one another. In addition, I think WOC try to emulate success that is defined within their company walls, and it causes some of them to keep a distance from women who look like them in favor of aligning themselves with their white male leaders and colleagues.

In 2018, research published in the journal *Development and Learning in Organizations* found that, when women undermine the credibility or status of other female colleagues, queen bee syndrome occurs. The research shows that senior women reinforce rather than dismiss the ideas of patriarchy and gender hierarchy. They assimilate by adjusting their leadership style to fit the masculine organization culture. These senior women distance themselves from less-tenured women, but they also don't serve as role models and may in fact stand in the way of other women progressing up the ranks. I think some of these findings can apply to what is happening among women of color.

Women of color have been on our own in these business structures for so long, it can be hard to build trust, even with other WOC. But if we are going to change the system, we have to start trusting one another. We need to see ourselves and our struggles in other WOC so that we can build compassion, alignment, and power for change.

We need to find ways to actively show solidarity with other WOC, regardless of their race or ethnicity. I have heard countless stories from Black women who feel that other WOC don't support or make space for them, or Latinas from one part of the world who don't support Latinas from other countries, or how WOC born in the United States look at WOC from their ancestral homeland with hostility.

One retired Asian leader in her seventies shared that she benefited from how some WOC—specifically Black women—did not conform. She shared she was more formal, and she tried harder when they would not play the game. But looking back, she wonders if she did those other women a disservice by conforming more because of her immigrant upbringing and need for success. She went on to add, "When you come to this country from other countries where there is struggle for food or basic necessities, you take every opportunity as a window. You climb through." She said there were a lot of people depending on her, so she didn't speak up or complain at work. "But now," she said, "ever since we started having more discussions on white supremacy, I question if that was worth it, and if I hurt other women of color along the way." She then dropped her voice and got noticeably quiet. The gravity of what she was saying seemed to dawn on her as she asked, "When those other women weren't conforming, it wasn't because they were angry, was it?"

We need to talk about how anti-Black racism and rhetoric shows up in our Asian and immigrant communities and how it is carried to work. As women of color, we need to do a better job of understanding that we are in this together. We can make it as a group, and we can make workplaces better for all. But we are going to have to pivot to more collective thinking. We need to admit the shared obstacles we have between one another, and the places where our history is different. And then we need to learn from one another, educate ourselves, and work with one another so we can all rise. Konda believes that if we come together and stop working against one another, we can have real power in corporate structures, because, as she says, "We are the backbone of the work that happens in our society."

And we need our white sisters to stand with us. Many of the women of color I spoke with said that white women in corporate spaces had been their biggest obstacles to rising. One woman shared a sentiment that was echoed by a few others: "White women not only didn't help me, white women went out of their way to stop me from getting my last promotion."

"We are all fish living in water that has been created by men, and it is the biggest catch-22," says Elizabeth Lesser. She strongly believes that we need to change the structure while living in it, even though it can be hard to make change when people are just trying to survive. "We are stronger in numbers and the group helps us not feel insane," says Elizabeth. "We need the larger group to know we are not crazy. And to confirm that we have been gaslit from Eve forward."

We need to join together in fighting the patriarchy within Corporate America. In addition to supporting us, women of color also need white women to understand that our experiences are different from their own because they have privilege by proxy of their whiteness. This creates a completely different experience for white women as they rise. It also creates a setup and a psychology where white women inherently trust the system, while women of color, especially Black women, do not. Therefore, white women will sometimes dismiss our experiences and the extra challenges we face in corporate settings.

White women need to work through their shame about enslavement, white supremacy, and their history of privilege. They need to gather and have their own discussions in order to work through the issues that hold them back from pushing against patriarchy. They need to agree to put their energy toward helping us rather than maintaining the status quo, so that we can all move toward and create what comes next.

In some ways, the process to truly come together as women in corporate settings will also require some form of shedding. But unlike the internal shame and anger we as WOC must shed, this shedding is about checking our behaviors and how we have shown up in the past. We need to acknowledge that we haven't always helped one another in corporate spaces. We need to see the ways we have been pitted against one another. As Elizabeth Lesser says, "We need to wash ourselves with a wave of forgiveness that we have not been as generous to our sisters as we could have been. It is understandable we aren't supportive of one another because the environment has been set up to be a zero-sum game." We need to forgive ourselves because, until now, we have not experienced a full model of sisterhood at work.

So, what can you do now? How do we teach ourselves to create stronger bonds across women of color?

FIRST, DON'T BUY INTO TOKENISM.

Don't accept that there is limited space at the top for WOC. Nina, who recently assumed a new senior role at a food service company, told me there are eleven other women at her level and above, and it seems as though they see her as competition. She feels like she has a target on her back.

What if Nina took the women to dinner somewhere off-site and suggested, "Let's help one another and try to create a new narrative that doesn't expect women to be limited to twelve seats"? How much more powerful would they be if the women supported one another instead of seeing it as a game of corporate *Survivor*?

SECOND, DON'T FORGET WHO YOU ARE AS YOU RISE.

The women I met shared stories about how, when women made it to the top of their companies, most didn't want to be seen as focusing on women's issues or inclusion topics.

One woman shared how her white female CEO tried so hard to avoid discussions around gender because she wanted to be graded as a great CEO, not as a female CEO. In an attempt to do so, she was very conscious of not wanting to promote or appear to be favoring women around her and elected to have an all-male senior leadership team. This attempt to not appear partial actually created significant bias and hurt many of the women around and behind her.

We need to work together as women in Corporate America to create what comes next. By banding together to address the patriarchy, the racism, and the sexism in our workspaces, we can try to ensure that the hive will not collapse.

THE OTHER BEES IN THE HIVE

Women of color need to work with men of color, too. My conversations have confirmed that men of color have a different experience at work. Jason, a Black male recruiter, knows that men don't have the double challenge of patriarchy and racism, and that sexual harassment happens much less frequently for men of color while it is a continual challenge for WOC. He believes that women of color have greater challenges around feeling safe in work settings because we are not often taught to fight back. He also thinks women of color have bigger cultural expectations to overcome because of what is demanded of us at home.

At the same time, he shared that we all have some overlapping challenges around stereotypes. Jason thinks deeply about how to present himself to people during an initial meeting so he is "relatable and not threatening." He also wonders if it's harder for men of color to find community and to talk about hurdles because they are taught to push forward and not share their feelings and frustrations.

Men of color also need to know that they sometimes hurt WOC at work, too, and we need that to change. A number of Asian women shared with me that they find it hardest to work with men of their same ethnicity because men will ignore them, talk over them, and refuse to take direction from them. These men will bring forth some of the traditional expectations of their culture. We need men of color to stop perpetuating messages they have been taught by the white male power structure.

Finally, white male leaders should understand that women of color's desire to change the system does not come at their expense. White male leaders have to do the work to understand how they are showing up, how to make space for others, and how it is their responsibility to help influence what comes next.

The challenges and obstacles for women of color are their problem, too, and white men need to participate and move away from the sidelines. My favorite example is a white male CEO who has a new rule that he won't attend a meeting without women and people of color present. He will walk into a room and walk out immediately if he doesn't see more than all-white male faces.

COMMUNITY CREATES POSSIBILITY

Ever since the 2016 election and Hillary Clinton's loss, women have been gathering with new fortitude and vigor. Her loss awakened a new urgency to come together to create change. The number of women at the Women's March, the number of women who joined and shared stories through Facebook's Pantsuit Nation group, and the number of women who ran for office showed us that the election results were sparking renewed conversations around women's rights, equal rights, and how we can work together as women to have greater breakthroughs. The election inspired further discussions about women and power, and now I believe it is time to talk about the next stage of that movement: women of color and power.

The events of 2020 created many new ways of thinking, especially around the fragility of our world. We appreciate the need for community and well-being more. We are finally willing to question the underpinnings of Corporate America, asking what comes next and what we can do to make it better. When we come together and support one another, we're stronger and more effective in changing these paradigms.

Gathering as women is something WOC cultures have done for centuries. Women gathered whether they were in tribes, villages, or small towns. It is how they found community, connection, and voice. Regional Chief Kluane Adamek shared that Indigenous women would gather because knowledge sharing wasn't passed through the written form; it was passed through verbal stories. She says matriarchs share their power through gathering and storytelling, and they show up for others, not just for themselves. Community is for the greater good, and there is a growing opportunity and a responsibility to participate in it.

"Creating more ways and places to share stories can change structures," says the chief. "By telling stories, we start to develop new frameworks." Sharing stories helps us see through the rules we have been taught and provides the awareness and foundation to change them. We need to gather and share new narratives so that we can see what is possible.

When I asked the women I met about movements by women that inspired them, they talked about Mothers Against Drunk Driving (MADD) and #MeToo. They explained that although these were movements created by women, the cause was something greater than themselves. These movements had a collective power and strength to create change, born from purpose and mission. I have also seen firsthand a number of new models where women of color are creating community to fight delusions, shed negative beliefs, and create change.

My former classmate Reshma Saujani created Girls Who Code to address the lack of girls in coding classes and the technology field. She saw a problem and decided to do something about it using a strategy I had never seen before. She spoke with corporate leaders to get volunteers, summer camp spaces, and funds. By having her programs take place on company premises, she helped companies feel invested in their pipeline in a very intimate and real way. Employers and employees alike wanted to be part of her movement and support her cause. She started with a handful of girls and a borrowed work space, but it became a movement with tens of thousands of girls trained with coding skills.

Reshma says many early factors, like confidence and fear of failure, create a situation where girls opt out of coding. By changing the structure and the factors that deter girls, she not only helped them learn how to code, but she also taught them to be brave and find their power. By creating a curriculum based in case studies and stories about girls, finding women instructors and role models, and helping these students create community, she learned that more of the girls gained the confidence to pursue careers in STEM. She created a sisterhood among her participants as well as a group mentality of togetherness, even if they were alone when they went back to their respective schools.

My sister, Roopa Purushothaman, created Avasara in Pune, India, to help girls become the next generation of women leaders. In a country where leadership is primarily taught to boys, the school uses a unique leadership curriculum to help Indian girls realize they, too, can lead and make a difference in their community and country. For Roopa, the idea started as a result of a high school project and a summer in India. She spent ten years buying land, building her model, raising money, and finding her first class of students. In 2020, Avasara graduated the first girls from the school.

As a founding board member, I met the girls when they first arrived on campus. Some of them struggled with their words and fought to make eye contact. Many of them were combating generations of patriarchy that suggested girls in India should not be raised and educated to be leaders and, if there was money to spare in a family, it should go to a boy child. But in four years, we have seen the girls change what they believe is possible for themselves and break through delusions that have stood for centuries. Some of the first graduates matriculated to college in 2021, with a few studying abroad in the United States and Canada.

What I find the most fascinating is that Avasara didn't just change the girls, it also transformed their parents and their community. There were huge doubts regarding what was possible when the school was built. As the girls kept learning, the parents started to believe anything was achievable for them. Even the most difficult parents, who were against sending their girls to an unfamiliar place, grew from this and praised how their girls had found their voices.

There are also white women working to combat patriarchy and forming more inclusive spaces that change outdated structures and create opportunities for both men and women. After being entrenched in the publishing industry for years, my writing coach, Suzanne Kingsbury, created the Gateless Academy. Her organization teaches writing in a new way that fosters supportive feedback rather than the rejection, rivalry, and feelings of inadequacy the publishing industry can breed in writers. She felt the industry was too masculine and layered with patriarchy. She wanted to create an approach to working with writers that fostered creativity and community.

Using Buddhist philosophies, writers work in groups so they can limit their inner critical voices. This approach results in writers gaining confidence and releasing their work into the world. She started with just a handful of writers, and now thousands of writers have been trained in her approach and eighty teachers around the world are mentoring in the Gateless method. Suzanne saw challenges in the writing process and created a methodology that combats the norms and isolation in the author process. She has created a community of authors who share tools, resources, and their networks. Since she could not change the publishing structure, she created a parallel structure that supports her writers in a warm and validating way. Her process helped me believe I could write my own book, and you are reading the result.

Ann Shoket, founder of New Power Media, author of *The Big Life*, and former editor in chief of *Seventeen*, also sees that power, collaboration, and community are changing. When she left her editor role, she started organizing the next generation of young people to talk about ambition and redefining power and success. She hosted dozens of dinners where she listened to what young people wanted for their lives.

Ann believes the next generation is changing what it means to be powerful. They don't believe power is tied to privilege and hierarchy, and they value transparency and diversity of all kinds. They want to work together, they aren't interested in trading their lives for their ambition, and they want freedom from the office structure. Most of all, they want their lives to have meaning. The young people she brought together have become helpful resources for each other, and they formed a community of support even after the dinners were over.

All these amazing movements started with an idea, a leader, and participants who were willing to try something new. They were founded to address some type of structural delusion, norm, or tradition. In the same way, I believe change is possible in our workplaces if we come together in community to create it.

Gathering women of color together and starting nFormation has also changed me as a person. A few years ago, as I was deciding whether to leave my corporate role, I realized I had been so busy climbing that I hadn't invested in building my own community. I didn't have a close network of people or relationships outside of work to call on, and especially not a group of women of color I could confide in. Former CNN news anchor Brooke Baldwin talks about this idea in her book, *Huddle: How Women Unlock Their Collective Power.* She says that even though huddles are productive and create conditions for change and progress, they don't always have to be about pushing forward. Sometimes, she says, "They are a space where women can simply bear witness to each other, or quietly sustain each other's very survival."

I now have an amazing circle of friends, including women I speak about in the book, like Rha Goddess, Tara-Nicholle Nelson, and Claudia Chan, who I will introduce in Chapter 9. Elizabeth Lesser calls this "womance": women you trust, tell your truth to, and share your heart with. These women have become more than sounding boards about work or personal-life issues. They are my community, my family. We support one another as the system throws us curveballs. For years, I heard that I needed to find female friends and sisterhood, and I ended up finding more than I ever could have imagined. These women helped me realize there was a bigger world beyond the corporate identity I knew. They helped introduce me to women who had left their old lives behind. They helped me through my loneliness and all my self-doubt, and they inspired me to construct a new life for myself.

Now I feel that, alongside them, I can push on intractable paradigms and change how we do business. To feel supported, we need community. We need community to heal. We need community to be seen, heard, and witnessed. We need community and the compassion it can unleash to address the challenges we see in the business world and beyond it. Once you find the power of "me" and find your sisters in the power of "we," anything is possible.

CHAPTER 8

HOW TO PLAY THE GAME WHILE YOU CHANGE THE GAME

Deb Elam, the former chief diversity and inclusion officer at GE, and the first Black woman to be a corporate officer in the C-suite at the company, is one of our advisors at nFormation. When I asked her about making changes within corporate structures and playing the game, she told me her goal at GE was getting to the top and opening the door for others around her. She paraphrased a line from *Hamilton*: "To play in the game, you have to stay in the game, and to win in the game, you have to have skin in the game." She believes to make change we need more WOC at senior levels of companies because, she says, "No game has been changed from the outside alone." We have to make change from the outside *and* from the inside. Reverend Al Sharpton once told her, "I don't need you outside marching with us, I need you in the boardroom explaining why we are marching." Deb suggests that there will be different roles each of us has to play, like the ones we talked about in Chapter 5, but most of all, if we want to make lasting change, we need more of us working together with the power of "we" that we talked about in the previous chapter.

As I wrote this book, I was surprised by the number of senior women who told me that I should focus on *how* to play the game. They felt that as women of color, no one had taught them to play the game, and we needed more instruction on understanding the structure and learning how to make the most of it.

There are so many books about getting ahead that I initially didn't even want to write this chapter. I didn't want to be accused of just showing women how to manipulate existing power structures instead of bringing in new leadership ideas. I've always felt if you acknowledged there was a game to be played and chose to play it, somehow, you'd succumb to the construct and cede personal power.

But Kia, the founder of Triangle Investigations who was introduced in Chapter 6, shared something I had heard before but never really taken in. "We need to play the game and run a new game at the same time," she said. "It is the only way we can get ahead. The game wasn't created for us, and I know you want to change it, but let's show the women how to play the existing game while we all lean into changing it." She's right.

This chapter provides an overview and general advice for how women of color at each step of their career can navigate and survive corporate spaces. That is what made my initial gatherings with WOC so special. It was like being in a secret conversation about what really happens in Corporate America, and learning lessons on how to stack the deck in our favor as we climb.

Early 2021 data shared by Catalyst suggests that if women are going to stay in corporate structures, we need a more level playing field. In the United States, women of color comprise:

- 18 percent of entry-level positions in corporations.
- Less than 4 percent of current C-suite roles.

And as we all know, women of color make significantly less than their white male counterparts:

- Asian women earn ninety cents to every dollar earned by a white man.
- Black women earn sixty-three cents to each dollar earned by a white man.
- Latinx women earn fifty-five cents to each dollar earned by a white man.

In order to change this, we first need to understand the various phases we go through in our work lives, and the rules and expectations of each stage. The different phases for women at work are *Learn*, *Earn*, and *Return*. At each of these phases, the rules and expectations will shift. This is especially true for WOC.

When we enter the workforce, we are focused on the Learn phase—learning the rules of our workplaces, learning how to do our jobs, and learning how to gain skills and advance in our careers. Around years seven to ten, we hit the Earn phase. We have hit our stride; we may have worked at two to three companies by now, acquired a set of skills, and moved from individual contributions to managing others. Finally, we graduate to the Return phase, which is about legacy and contribution. It is when most of us start looking back at our careers, joining boards, and asking about how we can make a bigger impact.

I could try to tell you the rules for each stage and give you a million pieces of advice for how to navigate them, but I can't possibly cover everything. Instead, this chapter will give you suggestions to adapt to common precarious situations you will face at each stage. You can use the tools and resources other women used to navigate almost all the challenges you may face at various stages of your own career.

LEARNING POWER MOVES

In doing research for this book, I gathered groups of less-tenured women who were early in their Learn phase. These women went into their corporate roles knowing they wanted the experience of working at a large company, but that they would not stay long. They went in with an "extract value" mentality to learn as much as they could to eventually start their own companies. They said things like "We know the system isn't set up for us" and "I don't even flinch when something happens anymore. I'm just here to soak up what I can, like a sponge."

The Learn phase is about getting used to the norms of the culture you are in, understanding what you are good at, and proving yourself. But for WOC there is also a large focus on fitting in and deciding how much to assimilate. During this phase, you may decide some company norms, like dressing in a particular way, are trade-offs worth accommodating. Other requests may be bigger. You will have to choose if you give in. You will learn there is no hard line. You will be faced with a constant set of active questions about whether you are giving up important parts of yourself. Start with "How does this make me feel?" when trying to figure out if a request is too far up the assimilation scale.

Karen Boykin-Towns spent twenty-two years at Pfizer, moving up the company from legislative affairs to chief diversity officer. Early in her career, she had to travel for work but had a young daughter at home, so when the meetings were done, she would rush to her hotel room to call home. Others would go to the bar, but she would stay in her room to relax and prepare for the subsequent day. She was content knowing she was missing the stories that would be talked about the next day. But her husband is a politician, and he told her, "You need to go to the bar. Even if you don't drink, that is where you will learn things that they won't talk about at the meetings, and you will develop deeper relationships." Knowing he was right, she started doing that.

"In some ways, looking back, it was a form of conforming," she said. "I was tired, a new mom, and I really didn't want to socialize. I wanted to rest." But attending the events helped her make strong relationships, and others got comfortable with her and learned more about who she was. Karen says it was important to share her life in these social ways. Her rise as a Black executive is a great example of how women learn, navigate, and gain help along the way.

There are many rules about conforming and assimilation when you are just starting out in the Learn phase. Some of them are straightforward, some contradict one another, and others are nuanced. The early basics include:

- Gaining confidence and skills.
- Doing good work and understanding what is expected to stand out at your company.
- Finding the right role models and mentors to help you get ahead.
- Learning that information is power.
- Getting along with others and appearing collaborative.

All of these are necessary, but they can be hard. At this stage, we are fighting many of the stereotypes around being a woman of color, and the bias people have around what a leader looks like. Robin Ely, a professor at Harvard Business School, told me that even early on, WOC can get caught up trying to manage other people's perceptions about them. Expending extra energy and effort to manage beliefs that are often based on negative stereotypes can be "soul crushing," depending on where you work, who you work with, and how you present in the world. The impact of this challenge can vary and be immense.

A WOC LEARNING PLAYBOOK

Lesson 1: Learn through osmosis.

Try to understand the norms and the written and unwritten rules about how things work in your company. Understand how people move up in the organization and what type of behaviors are rewarded. Learn how promotions and ratings work at your company. Every company has unwritten rules on what truly matters.

This information might get passed to you after hours, like much of it was for Karen at the bar. Many WOC wait for invites to these gatherings. Some WOC shared that they assumed gatherings after work were private, and they didn't even try to attend. My advice is that WOC who are trying to make a name for themselves get comfortable showing up, even without permission or an invite.

Lesson 2: Learn through mentors.

We are taught to find mentors who will help open our paths, and we need to listen to advice about how the company works from these trusted sources. In her Learn phase, Karen was lucky to have Brenda, a pioneering Black woman, take her under her wing.

Brenda explained to Karen that she was never going to "out-white a white person," and she needed to own who she was and what she brought to the table.

Karen pocketed that advice even if she didn't know how powerful the statement was or what to do with it early on. This instance is a good reminder that we will get advice, sometimes profound, but not know what to do with it at the time; that's okay. Tuck it away and take it out when the time is right and you are ready.

Lesson 3: Learn through "we."

In addition to mentors, advocates, and sponsors, it can make all the difference to have the "we," that group of peers that we mentioned in Chapter 7. Some women I met, especially younger women, are sharing their pay details and reviewing feedback openly with one another. Some have created Google Sheets where women input data about pay so everyone has access to that information. This kind of support ensures we get paid our worth, get reviewed, and advance fairly.

Other women turn to books, online courses, and even outside coaches to help them learn about the politics and practices within their companies. It is especially important that we get creative because, more often than not, the WOC I met didn't have others in their families who have navigated corporate structures. So, when they needed advice, many of their families could not help them.

During the Learn phase, many women are just trying to stay ahead of the curve. They're gaining the confidence to know that many of the insecurities they feel and the imposter syndrome they carry are less about them and more about the structures they are navigating. Use this time to shore up your defenses and your tool kit.

It's also important that you not be too tough on yourself in this phase. If we are asked to speak about our accomplishments, we need to do so with clarity and vigor. This was extremely hard for me to do, but I developed a technique I used for many years. I would share updates on my projects and highlight my team members' successes via email to my leaders. This was my way of boosting others while also telegraphing what was happening with my work, since that could sometimes feel uncomfortable or boastful.

EARNING POWER MOVES

During the Earn phase, many women of color fall out and leave corporate spaces because that move from individual contributor to being responsible for other team members can be hard for us. It can be uncomfortable to lead others formally, especially if we are shy or have been taught to be more reserved.

This is also the time when many WOC are hitting the phase of their lives where they are getting married and/or having kids. Many of us become responsible for taking on even more at home. Our parents are also growing older, and we may need to spend more time caring for them. This phase can last twenty-five-plus years for the women who stay in traditional work settings.

Karen shared another career-defining story from the Earn phase of her career. At the time, she was a senior director and being considered for promotion. For two years, her reviews were good, and she was on a visible project, but she wasn't moving to the next level.

She remembers getting a call from a C-suite executive just to check in. He asked if there was anything she needed. She said she was "fine." She remembers going home and telling her husband what happened, and he told her if she wasn't clearer that she deserved to be promoted, she would get lost in the fray. "They will stop asking you what you want if you don't ask for more," he said. (It is so true, and I want to meet Karen's husband, because he gives the best advice.) We need to know what we want when we are asked, because we don't get asked often, and we need to be willing to ask for more.

After the encouragement from her husband, Karen went back to the executive who called her. She asked for more direct feedback on her career and made a case for why she deserved to be a VP. She got both the feedback and a promotion.

I was similarly caught flat-footed midcareer. I was a new partner and the senior partner, who led my practice, called me to dinner. Twenty minutes into the meal, he asked me if I wanted to be CEO one day. He said leadership had talked and wanted to put me into succession planning processes, but there would be a series of roles I would need to take on to get ready over the next several years. I was deep into trying to reconcile my personal life. I had ended a long-term relationship and was worried I would be single forever. So when he shared his point of view, I froze.

I was caught off guard because I had never heard of this type of succession discussion happening this early on, and I didn't know if I wanted it. I had other priorities. Somewhere over the next fifteen minutes, I found a response that felt authentic to me and bought me some time. I told him that one day soon, I might want to get married and have kids, and if I did, I might want to make different choices. "If you can understand that," I told him, "then yes, I want to be considered until I don't want it anymore." He was bewildered by my honesty, and boy, did I hear about that answer for the next few years. But it still feels like an amazingly powerful answer that was true for me. You must find a way to authentically say what it is you want when asked.

During the Earn phase, you need to focus on:

- Building your reputation and executive presence.
- Finding opportunities to show leadership.
- Speaking up and being heard.
- Protecting your boundaries and well-being.

Many WOC find it hard to show their readiness and flex their executive presence because, as we talked about before, there is a bias in what people think leadership looks like, and the criteria for advancement is flawed. Recruiters and companies reward prior experience over other leadership qualities. They want us to have been in the role already. This results in the same people getting the same opportunities. They need to place value on the unique leadership qualities WOC bring to the table related to our life experience and our grit. Women who are a first or only have fought especially huge challenges to get to our seats within the workplace, which makes us stronger leaders than most. Many women I met were raising children on their own or were the first in their family to have a professional role or even a role outside of the home. That gumption and fortitude have to equate to strong leadership and management skills.

So many of the women I met shared that they got problematic and sometimes vague positive or negative feedback on their assertiveness or style. Some of the Asian women I met shared feedback about needing to be more assertive and stronger to be seen as leaders, while some Black women and Latinas were told to tone it down and be more accommodating. It can feel like there is a "just right" Goldilocks recipe that corporate is expecting WOC to cook, while never giving them the exact temperature on their recipe card.

A WOC EARNING PLAYBOOK

Lesson 1: Stand your ground and find your leadership style.

Data shows that more women, especially women of color, get stuck in middle management positions, otherwise known as the "frozen middle." While navigating the cultural expectations of our company, we need to work on our executive presence and find ways to be ourselves. We then need to be clear about what we want and seek promotions, like Karen did. In many cases, WOC do not get these stretch or high-profile opportunities because, as we spoke about in the first chapter on delusions, studies show managers and leaders tend to favor and give more chances to people who look like them.

We also need to find ways to be seen and heard. So many women I met shared experiences about being talked over and having their ideas stolen and repeated by the white people in the room. Mary Okoye, a Black judge turned lobbyist and consultant, says she has learned to call out when her ideas are stolen because it has happened throughout her career. "Remember you are supposed to be at that table, and you *are* at the table," she says. When people steal her ideas, she just repeats them again. "No anger," she says. "I just report the facts. I say things like, 'Isn't that the same idea I just suggested? How are you seeing it differently than what I just shared?'"

At the Earn phase, making sure we speak up and get credit for what we do will become even more important. Before this stage, some of the reputation building was done by others. At this stage we need to learn how to do more of this for ourselves and learn how to give and get feedback.

Lesson 2: Have enough sponsors or spies on the inside.

We also need to be clearer about what we want and need in this stage of our career. We may get thrown opportunities to ask for what we want, like Karen and I did, or we may have to find ways to make our desires clear. For many of the women I met, this is the hardest part of the career path, because we have been taught to be thankful and to not push for more.

Carla Harris, an executive at Morgan Stanley, has a great TED talk where she explains that the decision makers—those we may not work with directly, but who will be in the room when promotions, ratings, and bonuses are determined—matter more than we know. She explains that companies talk about meritocracy and fair processes, but people's views are ultimately subjective, and these leaders have great influence on an employee's review and rise. It is important to understand your "sponsors" and who has influence in your workplaces. As WOC, we may have to be more strategic and intentional about how we find and cultivate sponsors. Ask for help if necessary, and go out of your way to create interactions and connection opportunities.

Lesson 3: Set your boundaries.

The isolation at this stage can be high. One of the most surprising patterns I found was the level of additional isolation many WOC mothers felt at this point in their career. The active messages they receive—that it is not acceptable to get outside help for family matters like child care, cleaning, and cooking—are real in many Latin and Asian cultures. Many white colleagues can't relate, because for them, it is more acceptable to get outside help and pay for support. So some WOC mothers shared that they feel even more alone and isolated when they talk about challenges around their family responsibilities at work. Some have even opted out of parental resource groups at their companies for this reason.

I cannot stress enough how important it is to take care of yourself and set your boundaries during this phase. There is so much coming at you from work and from home, and it can often feel like you are failing on all fronts. The constant advice I heard from the women I spoke to was this: we need to take better care of ourselves, learn to rest, and find joy in the successes we have. Because in this phase, more often than not, women are exhausted and just trying to stay afloat as they rise.

RETURNING POWER MOVES

During the Return phase, some WOC may still be accelerating and rising, but we are starting to reach senior levels of our companies, preparing for boards, giving back, and questioning if the climb was worth the sacrifice.

As Karen made the move from Earn to Return and was promoted to VP, she requested 360-degree feedback from her peers and managers, and was told she needed to smile more, be more approachable, and share more of what she did on the weekends so that her peers could get more comfortable with her.

She wanted to get angry and suggest most of it was because she was a Black woman and simply different. But after praying about it, Karen realized this situation reminded her of the discussion years before when Brenda, her mentor, had told her she was never going to "out-white a white person." She needed to own who she was and what she brought to the table. Karen met with the CEO, a big supporter of hers, and shared the feedback she had gotten in an attempt to be proactive and be transparent about the concerns that were raised. He told her he wasn't worried about the feedback and simply said he'd been waiting for her to ask for more responsibility.

Karen says that at large companies, good work will only get you so far. Eventually every WOC will need a turbo boost, a push from someone senior. They use their star power to move you into the next level. He was her turbo boost and eventually offered her the chief diversity officer role for the company.

As one of the only Black women at the executive table, she raised issues about access that others could not speak to. She used her C-suite position to talk about access to medicine and underserved communities. She found a way to bring her whole self and what mattered to her to work. "We all have power," Karen says. "What it looks like may vary, and we get to decide how to use it and how to share it with others."

I love Karen's story for many reasons. She was always able to keep her personal passions core to her work, she found her power and voice, she was clear about what she deserved and asked for it, and she was finally able to be herself. She let her shoulders down, and when that happens, so many more things are possible for us.

In the end, after twenty-two years with Pfizer, Karen left to start her own consulting firm and focus on outside activities. She was recently reelected as the vice chair of the NAACP National Board of Directors. Her career beautifully weaves political activism with traditional corporate roles. She is one of the only women I met who was able to stay so heavily invested in civil rights issues and politics while navigating senior corporate life. She became uniquely positioned to be a growing voice on issues around systemic racism.

Women of color during the Return phase should focus on:

- Living your purpose.
- Defining your legacy and contributions.
- Opening doors for others.
- Leading in your own ways.

Deb, former chief diversity and inclusion officer at GE, has advice that follows nicely with what Karen shared about her rise. Deb says many Black women are taught to run faster and do more. She doesn't buy into this. Instead, she advises WOC to be savvy and strategically move ourselves into the game. She says we need to remind ourselves what we know, make sure we are on top of our game, figure out who the power players are, and do the meeting before the meeting so people cosign on our ideas. She says these extra tasks may not be fair, but they are effective.

Ultimately, Deb learned to put aside what other people thought. "Focus on your agenda," she says. "I'm not changing hearts and minds; I am moving my agenda." Early in her career, she wore suits and women's bow ties, mimicking men. She didn't display Black art or personal items so people could see she was "there for business." But over the years she was able to let go of some of her conforming. As she grew into her skin, she felt less constrained because her reputation became more established. She believes there is a level of conformity and there are norms we're still expected to follow, but she says, "What you have to decide is when you are being someone you are not. Nothing I did went against who I was or who I am."

Deb says it is especially important to decide what you share. People make decisions about what they think they know about you. She was intentional in sharing certain things about herself, and it helped her come off as open. That dichotomy on what to share is a big issue for women of color. "As a Black woman who got to 'rare air,' there is nowhere to hide," she says. "I have to show up and bring it. I don't get to fade into the background." She also stresses that as you get more visible, you should take time away from work to replenish.

She ended by advising all of us to develop external relationships so we have options if we need another job. In the end, Deb defines "power" as "having the ability to influence other ideas and actions to move the agenda forward." Now she works to help women advance, because, she says, "There are moves you can make that can be taught." And she believes that as we all get close to the top, we can open the door for others.

A WOC RETURNING PLAYBOOK

Lesson 1: Use your networks to help others.

We are often isolated from each other at our jobs and in the business world, so we rarely have the opportunity to share our experiences at work. Similar to what women did in the Earn phase, women I met in the Return phase mentioned they organized WOC networks in their companies to meet after hours or on weekends to discuss advancing through the company. Not seeing resources that worked for them, they organized and created networks themselves to help women who come after them. Other women are organizing their own board-ready sessions and asking recruiters to meet with them.

Lesson 2: Negotiate, negotiate, negotiate.

Jana Rich is the founder and CEO of the Rich Talent Group. She is a renowned recruiter and strong co-conspirator who specializes in recruiting diverse talent to leadership teams and boards at the most innovative and desirable companies in the US. She says more and more companies are committing to diversity, but are uncertain about how to effect change.

Jana and I had a long conversation about WOC negotiations at senior levels. She said the most obvious difference between WOC and other groups is our limited access to traditional networks and our lack of confidence in negotiation. "Women of color need to know what they are worth, be firm about their targets when they negotiate, and stand up for themselves." Jana said men will put out an ambitious ask and have good rationale and facts to support their request. They will have done their homework. They may not get the ask, but they are okay with that. Women of color often won't even make the ask or start the negotiation, and this is an area of huge opportunity.

Jana also believes that WOC may find false hope in new legislation that makes it harder for employers to ask about past pay. While the legislation is aimed at reducing the pay gap, the idea that we won't get asked about previous compensation doesn't make for a level playing field if women aren't bold enough to ask for what they want. She says that unlike the Saturn car brand and their no-haggle sales approach—where there is a set price for a car—finalizing a job offer is all about negotiations and it would be better for women of color to actively ask for what they need and want.

As we talked about in Chapter 5, many women of color get asked to take on additional roles or activities as they rise in their careers. Senior WOC are often specifically approached to take on culture-building and mentoring roles. Instead of saying just yes or no, there are ways to say yes and negotiate for other things that will help us in our careers. For example, some of the women I met accepted new committee assignments but asked for other highly coveted roles when they were done.

Lesson 3: Move from success to significance.

So many women I met are trying to move from a stance of "How do I have success?" to "How do I have significance?" In the Return phase, women are taking stock of how far they have come and where they want to leave their mark. There are more inner questions around purpose and impact, and that reflection can take up large periods of time and make women of color question their identity and worth.

Having significance means asking different questions about what you need and want as you explore opportunities. Most of the women I met who were grappling with these questions had made their mark in their first careers and were looking for new opportunities where they could give back, boost others, or profoundly influence the culture and agenda of an organization or company. But I also think we need to remind ourselves that it doesn't have to be some prestigious seat on a nonprofit board or picking up and moving to another country to work on poverty or lack of access to clean water. Sometimes just offering our advice to the WOC around us or helping a new WOC start-up can be more significant than we realize in shaping what comes next and helping other WOC avoid the pitfalls of racism in our structures.

I also think it is important to remember that the road for the first, few, and only is hard. The women who came before me did the best they could with what they knew and with the structures around them. Some tenured women I met were deeply questioning if their time and efforts have been worth the sacrifice. Maci, who we met in Chapter 1, is a public company CFO and a Black woman who can pass as white. When people would say racist remarks around her, she would disappear into her surroundings, not acknowledging that she was in fact a person of color. Now she regrets her actions. But she had always believed if she just put her head down, she would get to the top.

She didn't quite realize she was assimilating in order to seem white, and she didn't know that "certain protocols and business practices were actually racist." She also shared that she tone policed herself, so afraid of being perceived as an angry Black woman that she often played what she was saying over and over in her head. Now, she realizes, "I needed to tolerate white leaders, but I needed to do work to change the protocols." She seemed so sad and regretful as she shared her reflections. I think no matter where we are in our career paths, we need to be patient with ourselves and how we have learned, coped, and managed, so we can do better when we know better and feel strong enough to do so.

REWRITING THE RULES OF THE GAME

No matter what stage of your career you are at, we are in a moment in time where we need to be evaluating delusions more regularly. We should use our power to question the existing rules, and eventually that can lead to us rewriting more of them or even getting rid of some of them.

One of the women of color I met in early 2021 sat in a series of meetings to select a new member for her nonprofit board. Each candidate had twenty minutes to make their case. The white male candidates all went past their allotment, with one going thirty minutes over his time before he was stopped. The two women of color who interviewed were the only ones who stuck to the twenty-minute rule, and when the two WOC reviewers brought that up for consideration, none of the white members seemed bothered that the men had ignored the rules. White men often create their own rules while we try hard to follow the ones we are given. Even while we play the game, we need to give ourselves permission to reimagine and even rewrite the rules.

We also need to give ourselves permission to accept the wrath that may come with breaking those rules. We have been taught not to upset the white male leaders in charge, so that even if they offend us, we should kindly and patiently explain to them the error of their words and their ways. The women I met agreed there is an unspoken rule that you don't want to make the white men in charge upset or angry. One of the white male leaders I spoke with told me when women of color speak out, they often poke the lions and bears of the white men who want to act like bad behavior and discrimination don't happen. You will rattle their cages, and they may come back at you and try to put you in a cage. But then he smiled and said, "But I think it's time for you to rattle and rattle loud, because white men have been sleepwalking for a long time."

There will be times in your career where you will feel the pressure to conform and work within the lines, and it will be tempting to believe that you have to do that to get to the top. The civil rights activist Ella Baker famously said, "I didn't break the rules, but I challenged the rules." You need to understand the game and learn the rules, so you know when it benefits to play by them or if it is time to try and create your own.

CHAPTER 9

THE POWER OF DECIDING TO STAY OR TO GO

As I lay awake at three a.m. one night in late November 2019, I knew the email I was about to send my CEO would change my life. I was pulling the plug on my twenty-one-year career—no, scratch that, my entire identity. I was about to intractably change all that I'd worked so hard for, and it literally made me nauseous. I'd given up so much to climb the ladder, yet I knew deep inside it was time to leave.

I was trying to feel calm, but my husband had already quit his job to pursue his own start-up. That voice kept kicking in—the one that pretends to be full of reason, but deep down is really soaked in fear—saying, *How can you possibly lose the income and the title you fought so hard for? You need to stay and ride this out. Just be thankful you are earning money.*

That wouldn't be my first sleepless night. I would wind up not sleeping for weeks. I was trying to sort through the guilt I felt, walking away from the seat I had earned for WOC as the first Indian woman at the table. I wrote and rewrote my goodbye letter to people I had known for almost my entire adult life. I called it my work obituary. I had spent the last twenty years trying to get where I was, and once I was there, I didn't feel powerful, and I wasn't healthy or happy. I knew it was time to make a change and embrace the many opportunities that were in front of me.

Many women of color, especially in Corporate America, have scratches and bruises from climbing the ladder. We've worked so hard to stay on it that we don't have time to ask ourselves whether we *should* stay or whether it's time to go. We've fought so hard to get to the proverbial table, we can't even consider walking away. But we have more options than ever before, and deciding to stay or go is a new power we have available to us at work.

In my first year at my firm, a senior partner took my team to dinner. I sat next to him at a big, round table in a noisy steakhouse, and we chatted about what I had learned so far. He shared that the best advice he ever got was to know his worth. To understand your worth, he suggested, find a recruiter, or apply for a new job every few years so that you know your value in the external market. Doing so will also keep your interview skills fresh and help you to know you can make money outside of your current role—and maybe even more than you expected.

He explained that it was a way to be confident and secure that we were staying at our job for the right reasons. Even though he followed his own advice and looked externally, he always stayed where he was, but it helped renew his energy and his commitment to his role. I still remember him saying, "Choosing to stay will make you better at your job."

But I think we need to remind ourselves that, at all stages of our careers, we have the choice and the power to decide where we invest our time, energy, and creativity. We spend more hours working than many of us do sleeping or with our families. We should feel good about our choice to be there. If we don't, we should feel confident to explore other options whether we are in the Learn, Earn, or Return phase of our careers.

Research shows some employees are now changing employers every four years on average. Even prior to COVID-19, women of color were leaving far more often than their white colleagues. According to McKinsey's 2017 report "Women in the Workplace," Black women have the highest rate of attrition—18.2 percent vs. 15.4 percent for white women.

This is even more acute now. Corporate America is facing a crisis. In March 2021, nFormation conducted a study with Fairygodboss that suggested two-thirds of over eight hundred WOC surveyed were contemplating leaving their roles within the next twelve months. COVID-19 has caused renewed tension in the conversations between health and wealth, between purpose and paycheck. More people are questioning what their work is in the world. WOC are asking these questions with even more urgency and more often.

Data shows that the impacts of COVID-19 and the "shecession" were highest among groups of WOC, especially Latinx and Black women. There have always been discussions and questions about WOC leaving corporate spaces, and now we are on the precipice of a great reckoning. Almost every one of the women of color I met has had a serious conversation about leaving her corporate role in the last six months. Women of color are tired, feeling undervalued and unseen, and are ready to demand more—and at the same time, we are more sought after than ever before.

Minda Harts, author of *The Memo: What Women of Color Need*

to Know to Secure a Seat at the Table, says women of color want to leave but often have an internal conflict about their investment in the company. For example, they might think, *Staying in this position has been causing me a lot of health issues but is also paying off a majority of my student loans.* So women of color weigh the equation and stay. Minda explains that we need to remind ourselves we have the power, and we are assets to the companies we work for. Staying is adding to our legacies and our résumés, so we should reframe our focus. According to Minda, we need to ask ourselves "what it looks like to make work, work for you." We need to make sure we get to our finish line in a way that is benefiting us. "Don't let the company's why impact [your] why," she says.

Minda believes many WOC leave because the system doesn't invest in their success. "Companies are always dangling a carrot of a promotion," she says, "but it is always next year." She believes that we then stay because of loyalty, but the "goalpost keeps moving." At some point, she believes we need to take power back. And we need to ask our companies: If it's not going to happen this year, then when? That way, we get to decide with full information if we stay. She also says sometimes we stay too long, and "We need to adhere to those expiration dates."

Often women of color stay in tough situations because of cultural conditioning, but if we stay in a corporate culture that is not allowing us to live our truth, there is a big risk. As detailed in Chapter 4, women of color are getting sick both mentally and physically. While I am not asking all women of color to leave Corporate America in order to create a movement of change, we have to stay healthy and strong through all phases of our career. Women of color cannot risk mental and physical health to stay in damaging environments. Not all situations are salvageable, and only you will know when to exit.

If you are or have been in the position of deciding whether to stay or go, you know how daunting that can be. Michael Bush, CEO of Great Place to Work, shared some really pivotal advice that I think will be helpful for those in this situation and those who will reach this point eventually. His data, collected from thousands of yearly survey results and used to curate the annual *100 Best Companies to Work For* list in *Fortune*, shows that once a woman of color feels out of place in her company, it can be hard for her to go back to a happier time. If things don't improve in three years, he believes WOC should leave—because if it hasn't improved in three years, it won't improve, ever. He also said that when high-potential women leave, they will usually be just fine. They have learned to survive and thrive within corporate structures, and, by the time they leave, they have cultivated strong networks outside their companies.

Though we may not know it, when we ask the question of whether to stay or go, we are beginning to ask an even bigger question about whether the system and the culture we are in works for us.

I encourage WOC who may be unsure about staying or moving on to answer these six critical questions. They will make clear where you are in your career and if it is time for a change.

1. Do you feel like you have control over your destiny, or is your destiny in the hands of people who don't understand or value you?
2. Do your values align with those of your company?
3. Do you have a sense of safety and trust in the people you work with?
4. Is your pay commensurate with your contribution and performance?

5. Are you still growing and learning?
6. Is advancement accessible to you?

Looking back at my own process, I learned that in the end, the staying or going wasn't the real issue, though it felt like the biggest obstacle while I was deep in it. It ultimately became a question of who I wanted to be, and who I wanted to become. Asking yourself the question of whether to stay or go is a vehicle for discovering what matters to you.

EVALUATING WHETHER YOU STAY

Most of the women I meet ask me if I have a list they can use to decide if they are in the right place. There is no definitive answer because each woman's list will vary slightly based on who she is and what she needs at the stage of life she is in. I tell women they need to take a bird's-eye view and evaluate whether they are doing the work they were hired to do, if they are being rewarded fairly, and whether they can survive and eventually thrive within their companies.

But if you are choosing to stay when your gut may be telling you it is time to go, here are some general guidelines that you should consider.

KNOW YOUR WHY FOR STAYING

Women of color are set up to stay in work and relationships longer than others because we've been taught that we should appreciate our opportunities. But when we stay because we "should," without questioning that "should," we can get stuck and lose power. In toxic or tough work situations, WOC must deconstruct their reasons for staying and understand whether they are staying because it is the "right" thing to do, or because they have something to prove.

Valerie, a VP at a medical supply company in Miami, didn't love her job and hadn't for nearly seven years. But she kept her head down because she had a son she wanted to put through college. That was her "why." She'd told herself long ago that she would sacrifice until he was through college, and then she could figure out what she loved. The day he graduated, she quit.

KNOW YOUR MUST-HAVES IF YOU CONTINUE TO STAY

Even though Valerie stayed, she had her must-haves. She wanted to have greater flexibility over her hours so she could spend more time with her son, and she wanted to be paid her worth. She raised this topic during her yearly reviews and made clear her bottom line. When you are staying for specific reasons and have clear must-haves, it becomes even easier to ask for them. You have a stronger sense of overall freedom, so you can feel free to demand that your must-haves be met. If they aren't, you know you will walk away.

If WOC decide to stay in a difficult work situation because it is just not the right time to leave, we often thrive by establishing goals for the future. If we are in a role but not thriving, we should try to understand what needs to happen or what milestones we need to hit before we can leave. We may stay because we are biding time to learn a new skill. There may be financial obligations we need to address before we transition. Maybe we are the primary breadwinner for our immediate or extended family, and we can't leave because other options haven't opened up yet.

But if you are staying for years, trying hard to make things work for the promise of something better, you may want to root down and check your truths. As I was deciding what I wanted to do about my old job, I had to sort through the belief that I had sacrificed too much and could not leave, and that maybe I should stay because people at my level didn't leave. It took me more than three years to wrestle through this and finally leave corporate.

SET BOUNDARIES AND DON'T LET YOUR JOB OWN YOU

Edie, a Black woman in Washington, DC, was reaching her breaking point at her large services firm. She was constantly being second-guessed and having to remind her colleagues that she, too, had the requisite pedigree for her role. Edie became unmotivated and uninspired, which gave way to depression. She wondered why she was giving up so much of herself and her family time to be in an environment that wasn't supporting her well-being. After some soul-searching and realizing she didn't want to be in a work environment where the daily default was that she had to prove herself and submit her credentials before each and every meeting, she declined the partner track, realizing she didn't want to push hard her whole life. She chose to stay a senior manager, the level before partner, and is buying time to figure out what she really wants to do with her life.

For her book, *Work Pause Thrive: How to Pause for Parenthood Without Killing Your Career*, Lisen Stromberg—culture transformation consultant and one of my close co-conspirators—interviewed and surveyed hundreds of women like Edie who decided to take career "pauses." Her research revealed that the vast majority of these women did not want to pause their careers, but the systemic bias against motherhood and the lack of programs and policies to support mothers in the workforce forced them to downshift or leave the paid workforce completely.

Lisen believes that for far too long, women have tried to work in a system that was not built for them. "Their health suffers, their careers suffer, their family lives suffer, all because the system is built for one type of worker: white men with wives at home to care for the needs of the family," she says. And that burden is even heavier for women of color.

When I was trying to decide whether to leave corporate, I often felt I had only two options: to stay or to go. Looking back, I realized I could have stayed. But for me, doing so would have meant a new mind shift. I realized I could stay if I chose to do my job differently. If I stayed, I would have to create better boundaries and take better care of myself. I would have found a better perspective on where work fit into my overall life, and I would have said no to more things.

DON'T LOSE WHO YOU ARE AND WHAT MATTERS TO YOU

Not everyone has to leave their jobs to find purpose or impact or to follow a nontraditional path. Rani from Chapter 3 shared that, prior to starting work in corporate, she volunteered in India with Mother Teresa. Rani loved the work and the mission so much she thought she might stay and become a nun. But when Mother Teresa heard of Rani's plan, she told her that not everyone is supposed to follow in her footsteps with that life of service. Mother Teresa told her to get as much education as possible and make as much money as possible so that she could be the most useful to the people around her. She said that your head and your heart are not mutually exclusive.

Claudia Chan, one of my close friends and the creator of the SHE Summit program for women, WOC, and allies, helps create cultures that "awaken and enable individual impact, inclusionary behavior, and whole-life well-being." Her training framework is called the Whole-Life Leadership method, and it helps companies retain and unlock the best talent, while allowing employees to thrive as whole people, not just workers.

She encourages employees to reframe their career as their "vehicle for social impact" and find their areas of impact from where they are. She refers to this as being an "impact intrepeneur" or a corporate change agent, and trains executive leadership teams to create social movements that embrace and leverage employee side hustles for their platforms.

Claudia introduced me to Puja Rojas, who grew her side hustle after developing fibromyalgia, chronic fatigue, and rheumatoid arthritis when she was involved in a devastating car accident. She thought she might have to leave the tech company where she'd built her career. As the head of HR, she certainly knew what resources the company made available to people with chronic illness, but she was afraid of being branded as "sickly," so she hid her illness at work.

In an attempt to share her knowledge and process her condition, she found creative outlets to tell her story. She blogged about fibromyalgia, demystifying the chronic disease and writing about the challenges of working and living with it. About a year into her secret life as a blogger, Puja's company took notice of her writing and offered to move her blog to the company's website. Over time, her clients started to hear of the health challenges Puja overcame, and she actually grew her accounts in the health-care space because of it. In sharing her story, clients saw new connections to their products and clinical processes and asked for her to lead important accounts. As she shared her story, her persona grew in the company as a result. If she hadn't found a way to integrate her life and her work, Puja feels sure she would have left.

ALWAYS EXPLORE YOUR OPTIONS

As that senior partner told me when I was a young consultant, always know your options, because there is power in gaining information and knowing you have choices and are not stuck.

While thriving may not be possible for WOC in all situations—either because it's just not the right job for us or because of bias and racism—I think it should be something we aspire to or want for ourselves. While it may not be the right time to pursue your definition of thriving because of financial reasons or other obligations, there are still conditions we can ask for, plans we can develop, or mind shifts we can make to get to a better place. Once you decide what thriving looks like for you, you can move toward places, people, and actions that get you closer to that, whether it's in your current workplace or in an opportunity you have yet to discover.

WHAT TO KNOW WHEN YOU GO

Lara was one of the few women who qualified as a C-suite executive-in-training at her company, and she was excited about the opportunity. Then, without seeking it out, she was offered an interesting opportunity to become CEO of a new start-up. When she called me, she was deep into figuring out what she wanted to do. She felt she was on a fast track to becoming one of the division presidents at her global company. But the idea of being CEO now, in the field of her true passion, really called to her. We started talking, and I asked her what would make her stay.

Her answers struck a chord. Even though she had sponsors whispering in her ear that she was on the fast track, she felt intuitively that she didn't belong in her current company. Being a woman of color made her stick out with the senior leadership team, and she wasn't sure her immediate boss had full faith in her capabilities. She always felt like an outsider and wasn't sure staying was worth all the drama, angst, and even the risk of giving up this other opportunity, which was a sure shot at leadership. She actually broke down as she was sharing her story and her concerns.

As I got off the phone, I felt like she was leaning toward leaving to go lead the new start-up, but when I spoke with her a few months later, she said she'd stayed at her company, but that just ninety days after that, she'd found a different opportunity at a competitor. She landed an amazing VP-level role at a company with a clearer advancement track, more support for her passions and skills, and a more global culture with a much more diverse international leadership team. Sometimes the choice is more complex than either/or. Below are some things you want to consider if you are thinking of leaving so you can do it with the most potential for success in the long run.

EXIT IN THE RIGHT WAY

If you do decide to leave, it is important to leave on good terms. Many women shared stories about how their old company became one of their biggest clients when they went out on their own or even moved to another company. Creating closure and maintaining relationships where you can are smart things to do when you decide it is time to move on. Your network will remain your network, no matter where you go.

TAKE A PAUSE IF YOU CAN

Leaving can also mean picking a new path and making different career choices. Step away from some delusions about what you must do to find something you want to do. Sometimes we may need to exit our current situation while we find our next landing place. Though most people will advise us that we need to move from one place to another, we may just need time to rediscover what we want next. At times we end up in situations for longer than we expected, or sometimes we just haven't found our true calling or a job that feels like a career. Other times taking what one woman I met called a "radical sabbatical" can be the best plan.

I was able to take a substantial amount of time off to focus on my health before I decided I was going to leave and before I decided what I did next. Mostly, I focused on getting better, but I also spent time dreaming of what I wanted to do with my life if I started over. I found time to cook, walk outside, meet friends for dinner on a weekday, all things that had been harder when I traveled for my role. I homed in on what I enjoyed doing and what I wanted my life to look like.

Taking a sabbatical or even an extra-long vacation can feel like stagnant time, but it is time away that can give you the greatest clarity on what comes next. Many of us come from cultures where we don't pause because we didn't see people before us pausing. The women who were able to take a break between roles didn't regret it. They stepped away from their conditioning of the day-to-day and asked themselves questions about what they really wanted.

FULLY VET YOUR OPPORTUNITIES

Sometimes it doesn't take a career shift or a completely different role to find out what is next. It may be a lateral role or a position with better pay, flexibility, or benefits. But, in all these cases, it feels like we are in a moment in time when WOC are in great demand, and we should be careful about where we take our talents and our gifts. We have more power than ever in where we want to go next. Not every company will let us shine and set us up for success.

We need to get into a better habit of interviewing companies for "suitability" in the same way they assess us for "fit." Not all companies are WOC-friendly. Not all places invest in us as leaders. More than ever, the WOC I spoke with are being careful and cautious of what they are told and where they decide to go next. They are asking to meet more leaders before they make decisions and accept offers, and they are asking more questions. Some are using informal networks to speak with other WOC in their prospective companies before saying yes to a role.

Vernā Myers, who I spoke about in Chapter 1, was a freelance consultant with a strong personal brand she had cultivated for over twenty years before she began her role as VP, inclusion strategy, at Netflix. In fact, she had started her relationship with Netflix as a consultant, so she had the opportunity to meet employees at every level and from different roles and backgrounds. So when Netflix asked her to join them, she'd already had the chance to experience the culture and meet some of the people before she accepted the role.

From the beginning, she was clear with the CEO that she would only entertain the role if it had real teeth. She also realized that by taking on the job, she could directly help to shape the cultural narrative because she would be able to guide her colleagues, who decide what stories are being told and how representative and authentic those stories are. When she started, she spent the first three months interviewing every executive at Netflix to make sure she could build trust, identify gaps, and design a comprehensive strategy to address those gaps and influence change. After over twenty years of swearing she would never enter a major company, she did it, but on her own terms and in a culture that would help her flourish.

MANAGE YOUR STRESS

Leaving can be stressful. We spend a good part of our awake hours at our jobs. Leaving what you know, even if it is not an ideal job, can be hard. Starting somewhere new at any level of the organization puts us back in some aspects of the Learn phase we talked about in Chapter 8. It will take time to understand how a new culture functions, develop relationships, and learn politics. Most of the WOC I met found starting over stressful at first, but then it became rewarding. Many shared that it took six months to almost a full year to get their sea legs at their new companies and to know who to trust. The pressure and visibility of being a first, few, or only in a new company cannot be underestimated.

LEAN INTO YOUR NETWORK

Find ways to leverage your community by using resources to help you through the transition, including coaches and therapists. Listening to stories about how other people exited their jobs and pivoted their careers can also help.

Bozoma Saint John, the CMO of Netflix, talks about how difficult it is to transition into a new company, especially as a Black woman, on the *Better Together* podcast. She suggests calling on the "power of we," and recommends getting in touch with WOC at a new company before starting in a role to learn the lay of the land. Even with people you haven't met yet, she says, it's a way to become informed about the culture we are entering, and a way we can do business differently and also help each other as WOC.

LEAVING CORPORATE TO BE AN ENTREPRENEUR

In a reversal of Vernā's story, I have met more senior women of color in the past few years who have left companies to start their own endeavors than those who have stayed in Corporate America. So many women of color I met shared that they left corporate spaces to be themselves and enjoy their work, to own their schedules and realize their ambitions, and to find meaning and success.

The women I met are leaving to set up companies that operate with WOC-friendly practices and offer a unique point of view of the world.

One Black woman, a professor who has written about gender, race, and class in the business world, said it didn't surprise her that many WOC opted out of corporate. "Who wants to die in Corporate America?" she said. "We don't want to take over the plantation. We can see what it is doing to white men. We don't want that." She emphasized that we don't want to sacrifice ourselves; we want to follow our own dreams and define our own legacy.

As I spoke with women who have recently exited, they shared that the system on the outside isn't set up for us, either. It can be tough to raise money, to get venture capitalists to notice us, and to sit across from a table of men trying to sell yourself. But even though the new pressure of finding stability and a constant paycheck now falls squarely on them, many still relish in this new agency.

Less than twelve months into my own journey of being an entrepreneur, I have found it scary and liberating at the same time. I have found moments where I am in my full voice and also felt the pains of not having a steady paycheck and not having the safety net of a global brand to back me up. But in the end, I would not alter my decision to leave. Going out on my own has taught me the power of my own network and allowed me to focus on work that is deeply meaningful in the world right now. Here is some advice I collected and things I thought about when deciding to leave corporate to start my own business.

TAKE YOUR KNOWLEDGE WITH YOU

Tiffany Dufu, author of *Drop the Ball: Achieving More by Doing Less* and founder of The Cru, a peer coaching platform for women looking to accelerate their professional and personal growth, shared that many WOC, especially senior WOC, forget they are experts and seasoned professionals when they think of leaving their corporate roles to start their own endeavors. She reminded me that I was one of them, and that we often take our wisdom and skills for granted. Tiffany often coaches WOC who are plagued with doubt about whether they have what it takes to start a company because of the delusions that tell us we are dispensable and dependent on the larger organization we came from for success.

BE OPEN-MINDED ABOUT RISKS

Tiffany wants WOC—and especially more experienced women—who aim to be entrepreneurs to see that going out on their own with a set of products or services is actually a diversified business model, and less scary than putting all our eggs in the basket of a corporate career. She says, "When your livelihood isn't tied to one brand, you can find your voice." She believes that WOC "parse themselves and leave parts out to get ahead in corporate structures."

BUILD ZEBRAS, NOT UNICORNS

Tiffany's biggest advice is that WOC need to carry with them how powerful they are and know that this power can be used to create businesses and workplaces that operate in new ways. "We forget we have had to sacrifice to rise," she reminds us, "and we forget what we have left at the door when we enter corporate spaces." Tiffany's unique ideas and her proprietary platform helped her raise $1 million in funding in 2019.

Whether you are leaving to create a for-profit company, to solve a social issue, or to create a hybrid or benefit company that focuses on social change, there are ways to intentionally place ideas of inclusion and equity at the center of the equation. In the article

"Zebras Fix What Unicorns Break," the authors and cofounders of Zebras Unite talk about how the current venture capital structure rewards quantity over quality, and quick exits rather than sustainable growth. Venture capital chases after "unicorn" companies rather than supporting "zebra" businesses that repair, cultivate, and connect. But "zebras" are real, and they are profitable and can improve society. They are also mutualistic in that they protect and preserve one another.

As we leave our roles and look to create our own businesses, I hope more WOC venture out to solve meaningful problems and repair bias, inequity, and the barriers that make it hard for all people to rise and thrive. We can expand beyond shareholder value and the delusion of profit at any cost. We can build companies that place equal weight on culture and employee well-being. We can create for-profit ventures that are steeped in doing good in the world.

Lisa, who I introduced in Chapter 2 on shedding, is a shining example of exiting on her own terms. She left a highly coveted partnership to launch her own fashion brand, Gravitas. The mission behind her company is to show all women that they can look great and be themselves, whatever their shape and size. Her silhouettes in a full size range speak to this inclusive ethos. She is taking what she learned in Corporate America and turning it on its head by focusing her design and business principles on radical representation.

GROW YOUR SUPPORT NETWORKS

Julia Collins, the founder of Zume Pizza and now Planet FWD, is focused on health and the planet, and she has raised over $450 million in VC funding as a Black female founder. She says it can be hard as a WOC founder and that it is important to build community around you. Overinvest in these communities because you will need them. Be intentional about developing them and think about them during your leaving process.

EXERCISE SELF-CARE

As you are building your company, Julia says, practice radical self-care. Make sure you wrap yourself in "positive nurturing energy" and pay as much attention to your well-being as your fundraising. There will be breakdowns. It important to take care of yourself so you can avoid or minimize these breakdowns. You will feel alone, but if you have your community of "we," know your power, and practice self-care, you can get through it.

PLAY TO YOUR STRENGTHS

Julia shared that when she slowed down after getting pregnant, she realized there was even more available to her as far as resources and her own energy levels. She hired a coach who taught her to focus on her "unique power as Julia, to find it and name it." Hers was joy. Julia has the following advice for women starting businesses: "Look at yourself, your patterns, and the places where things flow to you." She says they may be signals of the things you are good at and the type of business you should start, and they also indicate where you may need outside help. "Lean into your differences. Use what comes easily to you as your assets." She realized that, rather than overworking herself, there was a new way to work as an entrepreneur, a way that was more in flow. That is when she found her greatest joy and her greatest success.

••

Weighing the pros and cons of staying or leaving can feel like the angst of a prisoner's dilemma, but there is power in owning the ability to decide. So many people go through their careers without realizing that, at every moment, they have a choice. Women of color in challenging situations need to move into a mindset that they don't need to tolerate systems that don't support them. Just like any relationship, only we will know whether to fish or cut bait. Sometimes we are ready to jump, and sometimes we are biding our time. Sometimes the workplace is too toxic to do anything but get out.

No matter what you decide—whether to stay and help create change from the inside, or to leave and make change from the outside—the power resides in you.

CHAPTER 10

THE NEW RULES OF POWER

One day a boy was walking with his grandfather and they came upon a large stone statue in the park. A big, burly man with flowing curly hair and huge muscles had his hands around the throat of a lion. After staring at the statue, the boy looked up at his grandfather and said, "I think this statue is all wrong. Shouldn't the lion be winning against the man? How is it even possible for the man to do this to the lion?" The grandfather simply replied, "Who made the statue? The person telling the story matters."

When I first heard this tale, it spoke to me, because there is power in being in control of our narrative. Those in power can construct more delusions if they own our stories. As WOC, it's time to take our power back.

Early in my career, I had a white male boss named Tom who would often tell me he played chess, and he'd point out how I was playing checkers. The first time he said it, I was confused but let it go. Over the year that I worked with Tom, I realized it was meant as a put-down. He was trying to tell me I didn't have the killer instinct to constantly be outthinking and outmaneuvering him, since he believed he was a strategic player who was always thinking multiple moves ahead. But what he didn't understand was that I wasn't playing chess *or* checkers. I didn't have a master plan. I wasn't trying to gain power that way.

Many women of color are stuck between knowing we see power differently, and at the same time being told that, in order to be successful, we have to wield power in an assertive, maybe even aggressive, way. In our capitalist society, where white men have gained advantage and wealth by exerting power over others, we have been taught that this is the path to success. In order to be successful, we are told we must be competitive. We have never been shown a different corporate playbook. As one woman I met described it, "the trappings of success" have been packaged so well that it's hard not to desire them or want to go after them.

Most of the WOC I met don't want to play the "old game" of power—a game that is based on the ideas of scarcity, competition, and winner takes all. One woman I met said, "2020 showed us that white men's definition of power is crumbling around us." Those ideals and models fell short as we faced challenges of a global health pandemic and renewed openness around concepts of equity and access. More people, and especially the WOC I met, aren't interested in gaining the type of power where only certain groups have access to resources. We see that power is found not only at the top of our companies. We each inherently have power within us.

Tom was all about gathering power at my expense, while I was gathering power for those around me. I thought we were on the same team, and I didn't understand his motivation or point of view. When we think of power, most of us have been taught to think about books like Machiavelli's *The Prince*, Sun Tzu's *The Art of*

War, or a more recent cult favorite, *The 48 Laws of Power* by Robert Greene. When I read these books, my eyes glaze over because the "kill or be killed" discussion feels completely disconnected from how I operate in the world. For me, going out on my own and immediately tumbling into COVID-19 and the evolving discussions around race have made it clear that we need a new rule book and different ways to lead. Power doesn't have to be bad—it can be used for good, and I want to create a version of power we can aspire to as women of color. This is part of the reason Rha and I started nFormation. We want to show women that the playbook we have been given is flawed, and it is time to write one that works for us.

Even as little girls, we encounter lessons that begin to teach us about power. Meghna Majmudar came to the United States at the age of five because her parents wanted to give their children a better life. Meghna studied hard and found herself participating in a city-wide spelling bee eighteen months after moving, but lost in the final round. The *Boston Globe* ran a story about the event, saying she would have been able to correctly spell the last word, "deceived," if she just remembered one of the golden spelling rules: "I before E except after C." The journalist wondered why Meghna missed a simple word. "We had just moved," Meghna told me, "and didn't know that rule."

The idea that there were rules was seared into Meghna's little-Brown-girl brain. She studied hard and during her senior year, when considering which college to go to, an advisor told her, "If you go to a school like Harvard, you will learn the rules of power and how it works." That was all he needed to say to get her to say yes.

And although Meghna went to class and learned some of the rules at Harvard, there weren't rule books to help her navigate social situations. She never quite felt like she belonged and found it hard to learn and adapt the rules of power as an immigrant.

When she graduated, she went into the corporate world, and those feelings stayed. She struggled to move past middle management because she made many social missteps and wasn't sure she was playing the game being played around her, and she often internalized her struggles. She never felt like her bosses saw her as their successor but as an immigrant woman. She now coaches "firsts and onlys" and especially immigrants about the idea that we are the permission needed to rewrite the rules.

If we are going to step into new power, we need new examples and new models. Professor Kimberlé Crenshaw, who coined the term "intersectionality" in 1989, said it well when she said, "It's not about supplication, it's about power. It's not about asking, it's about demanding. It's not about convincing those who are currently in power, it's about changing the very face of power itself."

WRITING A WOC-CENTERED DEFINITION OF POWER

Many WOC I met had read the typical books on power by Machiavelli and Greene, but few knew where to go to talk about leading and thinking differently. I want to suggest a new list of rules for the "game of power" that actually works for us. A list that allows us to rewrite power so more of us are interested in aspiring to wield it, instead of shying away from it. A list that stands as an alternative to the white heterosexual man's power list that came before. A list that represents new power. As I was researching traditional power, I often saw it talked about in four ways:

1. Power over—power that is based on domination and control
2. Power within—power related to self-worth

3. Power to/for—power that creates or achieves a result or goal
4. Power with—shared power that grows collaboratively and with others.

Have you seen MC Escher's black-and-white optical illusion paintings? In *Day and Night*, one of these famous prints, a flock of white birds are flying east and a flock of black birds are flying west. Many people will look at the picture and see only one set of birds, but the rare viewer will see both vantage points.

Women of color are these rare viewers; we are the only ones who are not confined to a limited perspective. Our wide lens and depth of perception can paint a new path forward that isn't constrained by what has been and what we only see around us. Women of color have the distinct advantage of being able to see through constructs that do not work (and never have). We can begin to write the new rules of power. Based on the ideas above, here are *our* first four rules:

RULE #1: "POWER OVER" IS OVER.

Our first rule must be that power over others is an old way of thinking about power, and we want a new one. We have talked about how, in contemporary culture, power has been portrayed as aggressive and assertive. That perspective isn't appealing or desirable to most WOC. When we emulate that definition, it's almost impossible for us to feel whole, aligned, or powerful.

Stacy Brown-Philpot, a Black executive and former CEO of TaskRabbit, shared a heartwarming story when I asked her how she defines power. She shared that her oldest daughter, Emma, was recently tasked with writing about a hero for Girl Scouts. Stacy assumed she would write about Vice President Kamala Harris or

Michelle Obama. Instead, Emma wrote about how Stacy is her hero because she helps her with math homework and is patient with her. For Stacy, this is a perfect example of her definition of new power. "Power isn't about how many people report to you, or a level or role," she says. "It's the things you do for the people that matter most and the ways we can show we care."

Stacy also shared that her family just got a new puppy, Fifi. To Fifi, Stacy is the alpha, not because she is the loudest or most aggressive, but because she most embodies stability and quietude within her family. Both of us talked about whether power and leadership should actually be defined by the presence and safety you bring. I love the idea that power needs to be coupled with safety, and that the leaders we gravitate toward should make us feel less guarded.

We have other models and examples of power available to us, but we don't always think of bringing them to work, and we should. If you can lead with heart and by example, creating safety for people around you, they will want to follow you out of loyalty and dedication, not fear. This type of power sticks and creates true followership where a culture of change is possible.

RULE #2: POWER WITHIN IS PARAMOUNT.

Our second rule of new power should be that we lead with aspects of ourselves that feel most powerful to us. Women of color need an alternative to the books that define power as being outside of ourselves. Power should be personal. Many WOC have been navigating in systems and processes that take power from them. The world doesn't see us as actual leaders and often doesn't see our inherent value. Instead, it pushes us to be different than who we actually are, pulling us from our true power.

True power lies in what we have inside. Iyanla Vanzant, an author, lawyer, and life coach who gained international attention through her work with Oprah, says, "Many of us think that our power is a tool or a weapon, but, in fact, power is a state of being, the way you see, hold and handle yourself in the world that then determines your experiences. The external world can only respond to the way you hold and handle yourself internally." She goes on to say, "Power is you. Power is standing up as your true self."

When I first met Kiki, she was in a palpable state of panic. She had risen fast in her consumer products company, but she felt strongly she didn't belong because she was the only woman of color—and therefore the only Asian woman—at her level. She felt frazzled because she was trying so hard to behave in ways that didn't come naturally to her and hiding many aspects of her home life.

A year later, when we met for lunch, she seemed like a different person. She was centered and powerful. This change came after getting feedback from her COO six months earlier. He told her she was good at strategy and getting the numbers, but no one knew who she was because she was "all work." As a result, she wasn't liked.

After getting silent, meditating, and taking a few weeks to sit with the feedback, Kiki took a "take me or leave me" attitude. She realized in her breakdown, as many of us do, that she could not be something other than herself. If she was going to fail, she wanted to at least be true to who she was and what she believed in. She stopped hiding her personal life and the things that made her different from the white male executives in the C-suite. She started talking about being an Asian woman and a mom. As a result of her new approachability and confidence, she was rewarded by the company with praise and new opportunities. But more important, she was relaxed, happy, and clear. She felt she had taken her power back.

You can also define personal power through what Robin Ely, a Harvard Business School professor, uses in her classes: the Reflected Best Self Exercise (RBSE). She asks students to use the RBSE to collect stories from twenty individuals who know them well. These stories showcase specific times when the individual experienced the student at their best, often during a challenging time. The tool helps highlight patterns where they positively impacted others. Instead of looking at external definitions, Professor Ely is asking her students to go inward to find their power.

Most women of color I met have found ways to armor up if they are going to stand up to the system. But remember that adapting to systems can take our health and our well-being, so we need to take on systems, not become them. We need to push back on issues, but not in ways that deplete us. We also can't beat the system with the system. Rha often says, "We need to show up differently if we want the world to be different around us." We need to wield our power in ways that feel congruent to how we think and live in the world. Maybe we don't meet force with force or rigidity with rigidity. When we show up centered, the system can surprise us and provide us with options and opportunities we haven't even imagined.

When speaking of power within, we can define this as within the self, but also within our own communities. MaryAnne Howland, founder and CEO of Ibis Communications and the Global Diversity Leadership Exchange, works with mission-driven companies and organizations committed to justice, equality, diversity, and inclusion, aka JEDI. As a single mom raising a son with cerebral palsy and ADHD, she felt she needed him to have more male influence and wisdom as he climbed into his teenage years.

She created what she calls the "Circle of Kings," which she wrote about in *Warrior Rising: How Four Men Helped a Boy on His*

Journey to Manhood. She brought together four successful men she respected, admired, and trusted to participate in his "Black Mitzvah", a rite of passage ceremony she created to commemorate their bond. They are working together to help shepherd her son into manhood, and to teach him how to realize his full potential as a proud Black man in today's world. She is not relying on the system as it is, but instead finding power within her own community. She is using it to reimagine power.

When I was in the corporate world, I put my center, or what I call my "locus of power," outside of myself. My identity and my power were shaped by outside validation and metrics of success. Now, when I think of how I feel powerful, it is inside. Power feels like alignment and reminds me that my strength resides within me. At its essence, this kind of power brings with it a liberation of choice that comes from seeing past the delusions we've been taught and understanding a new clarity that we get to create for ourselves. New power for women of color comes from knowing what matters to you and knowing who you are. Living your power means acting from this place.

RULE #3: POWER FOR GOOD IS STRONGER THAN POWER FOR SELF-GAIN.

In Corporate America, power is usually a rigid, predefined ideal based on dominance and exploitation for personal benefit. The more you take, the more you make. But for WOC, having power can't come with exploitation because we have historically been the ones being exploited. Most of the WOC I met believe that, as a group, we look at power differently as a result.

In their book *Power, for All: How It Really Works and Why It's Everyone's Business*, Julie Battilana and Tiziana Casciaro describe the "Three Pernicious Fallacies" of power as:

1. Believing power is a thing you possess.
2. Believing power is positional.
3. Believing power is dirty and acquiring it entails manipulation.

These fallacies suggest that we need to expand definitions to see that power isn't tied to seats at the top or authority. We all have power. But to ensure power for all, we have to engage in change.

Rha, my business partner, talks about power and fallacies in a similar way, suggesting the paradigm of power has been set up with three distorted notions: "Greed, which is focused on taking it all; dominance, which is focused on controlling it all; and exploitation, which is using it all." She says power, as it is taught to us, asks us to go after these goals by any means necessary. Just reading those statements may feel out of alignment to many WOC.

Technology is changing the ways people have access to knowledge and information. Participation in issues is rising because of social media. Because of this, the very essence of power is changing. Although wealth may still be accumulated by a smaller number of people, influence and access are growing in ways that can give more power to more people.

The old definitions of power are giving way to new values-based definitions of power. The women I met believe that going forward, power should be used for good. They described it as renewable, bold, and challenging. The women I met believe power for power's sake is passé. It should help us make a society better. One woman described this concept as taking power and turning it into significance. In this third rule, new power is less tied to the positions we hold, and more tied to what we value and how we stand up for the things that matter to us.

RULE #4: "POWER WITH" REIMAGINES THE CORPORATE TABLE. I am going to put the most attention on Rule 4 because I believe banding together as a collective is the most important thing we can do to create change.

Mary, the Black judge turned lobbyist and consultant from Chapter 8, says that when people decide that they've had enough, the system will change, because we have the power to change structures. She continues, "When one king dies, and a new one is put in his place, for centuries people have said, 'The king is dead, long live the king.' The people have the power to decide who and what is in charge, not the systems around us. We give the systems and the structure their power."

Structures are falling around us. Everyday norms, including our vocabulary, ways of doing business, and outdated societal constructs, are changing fast. Our meetings have already transformed from conference rooms to Zoom squares. The conversations we are allowing ourselves to have in the workplace are also moving quickly.

Perhaps the table as we knew it is no more. Maybe there no longer have to be set seats at the table. Perhaps it's time for corporate tables and boards to add multiple seats so more of us are at the table and our voices can be heard. States like California are already requiring change by mandating companies to include board members from underrepresented communities. As new perspectives arise, the seats of power need to change and move so that the static nature of the table evolves.

All this means there is an opportunity now to take apart the table that exists. We can keep the parts of the table we want to, while rebuilding anew. Nicole Anand is a political economist, a participatory designer, and an adjunct faculty member at Parsons School of Design, The New School. She is also leading systems initiatives at UNDP. I met her after she wrote a paper for *Stanford Social Innovation Review* on how the way we work causes assimilation among workers.

When I asked Nicole about the new table, she said, "I don't even want the table to be there. I think we need to create more spaces where we can invite women of color in to create new solutions." She believes that this is where the new power lives—in spaces where we are rethinking the tables of the past. She wants us to stop aspiring to a seat at the table and reimagine it as "a fire that we all stand around" where everyone is warmed, and everyone contributes by putting logs on it to keep it raging.

She's not the only woman of color who wants to redesign the table and its rigid rules. Ava DuVernay emphasized the importance of this in her 2019 *Glamour* Women of the Year Awards speech, saying, "I don't want a chair at the table, or even three, or even half [of the table] anymore. I want the table to be rebuilt. In my likeness. And in the likeness of others long forced out of the room."

Some of the women I met take the idea of remaking the table one step further by talking about needing to evolve capitalism. They believe capitalism is a form of sociopathy. What I heard women saying was that we need to take the individualism we prize in Corporate America and marry it with ideas our ancestors had about the collective good, where outcomes and maybe even business decisions can be tied to impact and future generations. If we are going to heal the game of power, it's time to take apart the delusions we have been taught about capitalism and the underpinnings holding it up, such as meritocracy, scarcity, and competition.

As we talked about in Chapter 1, these beliefs make us think there are limited seats and even fewer seats for WOC at existing tables. The setup of the old table assumes that a seat needs to be vacant in order for us to take it. There is an idea of inherent scarcity in the concept of the old table and a belief in some version of survival of the fittest where we must compete with one another for the rare seats.

We also need to go back and look at the origins of capitalism in the United States. "Racial capitalism" is a phrase coined by Cedric J. Robinson, a professor from the University of California in Santa Barbara, and he used it to talk about the process of extracting economic value from a person of a different race. Konda, who we met in Chapter 7, says, "Racial capitalism was built to be transactional. It was built on enslavement." She notes that we can't put the history of capitalism aside, bluntly saying, "Capitalism has failed, and we can't create a healthy system and society with a broken financial system." We need to get to the root of the problem, and her belief is that anything that has been created can be uncreated.

White men have historically changed the rules of the game when it suits them, making it even harder for us to succeed. It's why I want us to think about new power. Nathalie Molina Niño, who I also introduced in Chapter 7, gave me an example of a new paradigm. "We need to remove all the racist obstacles that keep Black and Brown people from things like buying homes and getting loans," Nathalie says. "Our buying power is over $3.9 trillion strong, larger than the GDP of nearly every nation on earth. If we concentrated our efforts, had each other's backs, changed how we banked, and pooled our resources, we could be the most powerful special-interest group in the world. That is the real power."

More than ever, it feels like we are in a time where power is collective, it has a greater meaning, and its very essence is changing. Just by being in a community and being witness to one another's stories, we will start to move the current to create change and harness the power we need to remake the table and the structures around it.

Gather in community with other women of color. Talk about what you each need to be supported and what we all need to make change. Talk about the cultures and the possibilities we could create if we didn't accept the system as it is. Continue to ask new questions. You can start with a few of these:

1. How does your background as a WOC impact you in the workplace and the world?
2. What racial or cultural background are you expected to shed as you show up at work?
3. What do you feel like you are being called to do as a WOC at work right now?
4. What are your models for new power?
5. What does it mean for WOC to have power?
6. What will be different for us and in the world as we claim and use our power?

As we work through corporate systems, we need to release the delusions of power and capitalism we have been taught and rebuild these structures so that they work for everyone. Arundhati Roy, in *War Talk*, has a powerful new vision for this. "The corporate revolution will collapse if we refuse to buy what they are selling. . . . They need us more than we need them. Another world is not only possible, she is on her way. On a quiet day, I can hear her breathing."

We are at a pivotal time and place. We must start or continue to question everything around us, even the largest and most established structures that seem to hold up our society. Our final rule is that together, we can remake anything that has come before.

WE ARE THE LIONS

Regional Chief Kluane Adamek's TED talk, "The legacy of matriarchs in the Yukon First Nations," was based on the idea that women in her tribe have always been powerful. She felt called to put her name forward to lead her people and have more women at the table. "I had to really look to myself to say, 'If not me, then who?'" In her talk she says, "There's so much that we can share with the world and that the world can learn from us as women. These are the challenges that we have for this future generation, and these are the challenges that we need to accept together. We need to give ourselves the permission to step into our own power."

When we talked, she shared an idea that stayed with me: we need to think of the world in a different way. "Look at a map," she said. "You may see a parcel. You will ask what the land is doing. Where is it sitting, what is the topography? That is a Western way of looking at a map. The Indigenous worldview would just be: the land is just there." The white man is seeing the map as a construct. The idea that two groups can look at the same thing and have fundamentally different perspectives, like the optical illusions I mentioned earlier, is at the core around new concepts of power. Power—especially in corporate spaces—is also a construct, and it's time we start to think of it differently. Yes, power is about resources, like the land, but power is also about how we see the past and our future, and how to take responsibility for what is next.

We are in a new transition space where old paradigms still have their roots, and new ways of thinking and operating in the world are just starting to germinate. This is the time where we need to build the foundations of new power. We need to grow it in ourselves, and we need to grow together. Our lineage, what we believe about ourselves, and our values and beliefs around power will shape what comes next.

Ciel, the healer I mentioned in Chapter 4, advised me to take a look at *Thus Spoke Zarathustra: A Book for All and None* by Nietzsche. Nietzsche talks about the evolution of people and says that most of them find themselves wanting to be a part of and following the herd. He says most people are happy to stay in that herd because it is easy and it is what we are expected to do.

Conversely, some people want to break away from the herd, into the journey of self-discovery, and these people turn into camels. They have "strong spirit," and they question why things are the way they are. They are ready to take on "the burden," like camels do, and the hard work of carrying what happens when they ask questions like what is next.

Eventually some camels will want to unburden themselves. They are willing to not only question their reality and the reality of everything around them, but they are ready to do something about it. This is when camels transform into lions. Lions are willing to question tradition, rules, and the status quo. They see rules have been established but never updated. They are fearless to step away from what they have been taught, to demand a new path and a new way. They are willing to say, "No more." They are fighters and will slay the "dragons," the existing rules and established way of doing things.

Fighting the delusions Nietzsche describes is true liberation and also hard work because it comes with a long history and centuries of conformity. We need to give ourselves permission to do things differently. I think many women of color are past the camel stage and have transcended to being lions, but instead of just saying no and questioning things, we must now build anew.

Through writing this book and having so many life-changing

conversations with other women of color, I've learned that we have a singular ability to listen to others, build bridges, understand historic challenges, and help people feel safe. We have the characteristics of lions. We are the brokers of new power.

A new era is beginning. As we become the growing majority of the educated workforce, we have an opportunity to be the change the world needs right now. We can come together and push against a system that was not constructed with our voices and truths in mind. Work is being redefined all around us, and the COVID-19 quarantine has made more of us question what work should look like and how much space it should take up in our lives. There is space for a new type of leader. A leader who places value on empathetic factors like well-being and quality of life. A leader who questions what was and pushes for what must be. With emerging leaders like Alexandria Ocasio-Cortez, Amanda Gorman, and Malala Yousafzai, we are just starting to see glimpses of what our legacy could look like.

At nFormation, Rha often concludes our initial meetings with women with the following quotation from Maya Angelou: "Courage is the most important of all the virtues, because without courage you can't practice any other virtue consistently. You can practice any virtue erratically, but nothing consistently without courage." For the first time in our history, it feels as if there is a recognition that our experiences, our truths, and how we walk in the world are unique and to date largely ignored. We all need the courage to see the world through a different lens than the current white majority, and to use that lens to create a new way of doing business—fairer, more just, and more equitable.

In some ways, because we have not fully come together as a group, and because we don't have a long history of visible role models, there are no predetermined ideas of who we can be or what we can accomplish. Our archetype hasn't been scripted. We have a blank slate to define our legacy. If we work together, I believe you and I—together with the hundreds of other women of color I met while writing this book, and our co-conspirators—will achieve our true potential and contribute on a scale the world has never seen before.

Our voices have been largely silenced and muted. Let's expel the delusions, tell our truths, and slay the dragons so we can wield our full power. Let's own our narratives and finally create a society that meets all of our needs now and into the future. This is our time and our moment.

Let's roar.

ACKNOWLEDGMENTS

I have loved the word "serendipity" ever since I heard it defined as fate meeting preparedness. I even used the word in my vows when I got married. This book has felt like a dance between serendipity and synchronicity. In some ways, years in the making, sprinkled with moments of divine intervention. When I needed the right expert, the right story, or the right advice, it would show up. It felt magical and it happened too many times to count.

There are so many people to thank that I cannot possibly list them all here. I want to start with Richard Pine and Eliza Rothstein, my agents from Inkwell. You agreed to this project and saw something bigger than I myself saw when we first met. You said yes in less than twenty-four hours and, as I am a new author and a woman of color, that is unique and rare. And my dream team from Harper Business. I wanted to work with a business imprint, and to have Hollis Heimbouch and Wendy Wong buy my book together and to have them both lean in as deeply as they did was such a gift as a new author. Thank you for believing in me. And Wendy, your voice, especially as a fellow woman of color, was invaluable.

Thank you, Suzanne Kingsbury. Without your help, I would not have had the confidence or the tools to put this work out in the world. This book would not have happened without you.

To all the women who shared their personal stories with me. The women I named and the women who shared their stories in confidence, thank you for trusting me. Thank you for being so generous with your time and your advice. Your desire to help lift the women who come after you is an inspiration and you made writing this book a joy.

To my early readers Nate Wong, Patrice Ford Lyn, Sukh Kundu, Noni Allwood, Jasmine Monfared, Lisen Stromberg, Ashley Davis, Professor Hannah Riley Bowles, Professor Valerie Purdie-Greenaway, and Professor Lakshmi Ramarajan, thank you for your advice and thoughts. You made this book better. Melissa Bannister, thank you for your fast edits and your keen eye. Elizabeth Michel, thank you for helping me validate sources and uncover important research.

Thank you to all my extended communities. To our nFormation women, my Aspen fellows, the REx group, Gateless writers, and my peers and students at the Harvard Kennedy School, thank you for always pushing me to go forward and encouraging me to believe that I could build new networks and relationships as I left the corporate world I knew so well.

To all the authors and experts who took time to meet with me. I heard so many stories about how writing this book might be a lonely experience. But it was far from it. People like Elizabeth Lesser, Dr. Margie Warrell, and Nathalie Molina Niño were so generous with their time and wisdom. Others like Professors Julie Battilana, Tiziana Casciaro, and Efrén O. Pérez shared their whole manuscripts with me. You showed me that we can do things differently if we choose to do them differently.

Thank you to my white male mentors and sponsors. Mike Fucci, John Powers, Bill Beyer, Schaffer Hilton, and Jerry Belson—you helped me climb the corporate ladder and you have stayed by my side as I chose a new path.

Thank you to my team, Sofiya Deva, Tammy Ari, Katie Wallace, Christina Welsh, Regina Dowdell, and Calgary Brown. Thank you for all your support and hard work on launching this book. I am so lucky to work with each of you. And to my extended team, people like Nishi Chaudhary, Willie Jackson, Katrina Frye, Dorie Clark, Elaine Bennett, Marie Incontrera, and Linda, I could not have done this without your support and counsel. And to my legal advisors Brian Melton and Debra Soltis, having access to your expertise kept me brave. Thank you especially to Tara, Claudia, and Reshma. Each of you helped me when I started this journey and stepped in with advice and counsel when I needed it.

Thank you to Carolyn Badila, Daniel, Dr. Rosenfeld, Lionel, Cat, P.C., Nand, Ciel, Dr. Lehman, and Dr. Shikhman. Without your help and healing, I would not have the physical and mental health to chase my dreams.

Thank you to Rha. Your coaching, partnership, and sisterhood is the backbone of this book. I could not have stepped into this next chapter without you and I am so excited by what we are creating together.

Thank you to my family, and especially my husband, Manoj Srinivasan. You supported me in leaving my comfortable old life and chasing my dreams in every way possible. I am lucky to have your motivation and your friendship every day. And thank you to my four-legged boys, especially Gunda. Always my biggest champion, you sat next to me and watched me write every word of this book. I miss you.

NOTES

Introduction

xiii we often endure microaggressions. "Microaggressions Are A Big Deal: How To Talk Them Out And When To Walk Away," *Life Kit*, NPR, June 9, 2020, https://www.npr.org/2020/06/08/872371063/microaggressions-are-a-big-deal-how-to-talk-them-out-and-when-to-walk-away.

xv Already, we generate $1 trillion as consumers. Cindy Pace, "How Women of Color Get to Senior Management," *Harvard Business Review*, August 31, 2018, https://hbr.org/2018/08/how-women-of-color-get-to-senior-management.

xvi Professor Efrén O. Pérez at UCLA suggests. Efrén Pérez, *Diversity's Child: People of Color and the Politics of Identity* (Chicago: University of Chicago Press, 2021); Efrén Pérez, "(Mis)Calculations, Psychological Mechanisms, and the Future Politics of People of Color," *The Journal of Race, Ethnicity, and Politics* (2020), https://doi.org/10.1017/rep.2020.37.

xvii Activist and author Alicia Garza. Alicia Garza, "A Conversation with Alicia Garza: What Is Needed to End Violence Against Girls and Women," audio interview by Jesenia Santana, Move to End Violence, August 11, 2016, 44:50, https://movetoendviolence.org/resources/conversation-alicia-garza-needed-end-violence-girls-women/.

Chapter 1: The Delusions That Hold Us Back

4 only 2 to 3 percent of Boeing engineers were women. Steve Wilhelm, "100 Years of Male Domination Not Yet Over at Boeing, New Book Says," *Puget Sound Business Journal*, October 16, 2015, https://

www.bizjournals.com/seattle/news/2015/10/16/100-years-of-male-domination-not-yet-over-at.html.

5 *Harvard Business Review* ran an article. Ruchika Tulshyan and Jodi-Ann Burey, "Stop Telling Women They Have Imposter Syndrome," *Harvard Business Review*, February 11, 2021, https://hbr.org/2021/02/stop-telling-women-they-have-imposter-syndrome.

6 91 percent white. Robert P. Jones, "Self-Segregation: Why It's So Hard for Whites to Understand Ferguson," *The Atlantic*, August 21, 2014, https://www.theatlantic.com/national/archive/2014/08/self-segregation-why-its-hard-for-whites-to-understand-ferguson/378928/.

7 white women and women of color combined. LeanIn.Org and McKinsey, "Women in the Workplace 2020," September 30, 2020, Exhibits 1 and 2, https://www.mckinsey.com/featured-insights/diversity-and-inclusion/women-in-the-workplace.

8 The well-known résumé study. Marianne Bertrand and Sendhil Mullainathan, "Are Emily and Greg More Employable Than Lakisha and Jamal? A Field Experiment on Labor Market Discrimination," *The American Economic Review* 94, no. 4 (2004): 991–1013.

8 In a 2016 study. Sonia K. Kang, Katherine A. DeCelles, András Tilcsik, and Sora Jun, "Whitened Résumés: Race and Self-Presentation in the Labor Market," *Administrative Science Quarterly* 61, no. 3 (September 2016): 469–502, https://www.jstor.org/stable/24758675.

9 many of us code-switch and adapt who we are. Courtney L. McCluney, Kathrina Robotham, Serenity Lee, Richard Smith, and Myles Durkee, "The Costs of Code-Switching," *Harvard Business Review*, November 15, 2019, https://hbr.org/2019/11/the-costs-of-codeswitching.

9 As Ella L. J. Edmondson Bell and Stella M. Nkomo. Ella L. J. Bell Smith published earlier in her career under the name Ella L. J. Edmondson Bell. I will refer to her throughout the text using the name relevant to the published work. Ella L. J. Edmondson Bell and Stella M. Nkomo, "Race in Organizations: Often Cloaked but Always Present," in *Race, Work, and Leadership: New Perspectives on the Black Experience*, eds. Laura Morgan Roberts, Anthony J. Mayo, and David A. Thomas (Boston: Harvard Business Press, 2019).

10 "Moving from Pet to Threat." Kecia M. Thomas, J. Johnson-Bailey,

R.E. Phelps, N.M. Tran, and L. Johnson, "Moving from Pet to Threat: Narratives of Professional Black Women," in *The Psychological Health of Women of Color: Intersections, Challenges, and Opportunities*, eds. L. Comas-Díaz and B. Green (Westport, CT: Praeger, 2013), 275–86

11 self-made and self-sufficient. Michael J. Sandel, *The Tyranny of Merit: What's Become of the Common Good?* (New York: Farrar, Straus and Giroux, 2020), 30.

12 Most of our white colleagues. Ella L. J. Bell Smith et al., "Easing Racial Tensions at Work," Center for Talent Innovation (now Coqual), 2017, 5; see Jason Fried, "Changes at Basecamp," April 26, 2021, https://world.hey.com/jason/changes-at-basecamp-7f32afc5; Casey Newton, "Breaking Camp," *The Verge*, April 27, 2021, https://www.theverge.com/2021/4/27/22406673/basecamp-political-speech-policy-controversy; and Casey Newton, "Inside the All-Hands Meeting That Led to a Third of Basecamp Employees Quitting," *The Verge*, May 3, 2021, https://www.theverge.com/2021/5/3/22418208/basecamp-all-hands-meeting-employee-resignations-buyouts-implosion.

12 Kareem Abdul-Jabbar wrote. Kareem Abdul-Jabbar, "Don't understand the protests? What you're seeing is people pushed to the edge," *Los Angeles Times*, May 30, 2020, https://www.latimes.com/opinion/story/2020-05-30/dont-understand-the-protests-what-youre-seeing-is-people-pushed-to-the-edge.

13 In June 2020, about a week after the murder of George Floyd. For a summary of corporate reactions to the Black Lives Matter protests, see Tracy Jan, Jena McGregor, Renae Merle, and Nitasha Tiku, "As Big Corporations Say 'Black Lives Matter,' Their Track Records Raise Skepticism," *Washington Post*, June 13, 2020, https://www.washingtonpost.com/business/2020/06/13/after-years-marginalizing-black-employees-customers-corporate-america-says-black-lives-matter/; Gayle Markovitz and Samantha Sault, "What Companies Are Doing to Fight Systemic Racism," World Economic Forum, June 24, 2020, https://www.weforum.org/agenda/2020/06/companies-fighting-systemic-racism-business-community-black-lives-matter/; and Ramon Laguarta, "PepsiCo CEO: 'Black Lives Matter, to Our Company and to Me.' What the Food and Beverage Giant Will Do Next," *Fortune*, July

16, 2020, https://fortune.com/2020/06/16/pepsi-ceo-ramon-laguarta-black-lives-matter-diversity-and-inclusion-systemic-racism-in-business.

14 "taboo or irrelevant in many business circles." Edmondson Bell and Nkomo, "Race in Organizations." See also Karen Brown, "The Fear Black Employees Carry," *Harvard Business Review*, April 30, 2021, https://hbr.org/2021/04/the-fear-black-employees-carry.

14 continuing to live in a neighborhood that is 71 percent white. William H. Frey, "Even as Metropolitan Areas Diversify, White Americans Still Live in Mostly White Neighborhoods," Brookings, March 23, 2020, https://www.brookings.edu/research/even-as-metropolitan-areas-diversify-white-americans-still-live-in-mostly-white-neighborhoods/.

15 According to LinkedIn data. Bruce Anderson, "Why the Head of Diversity Is the Job of the Moment," LinkedIn, September 2, 2020, https://business.linkedin.com/talent-solutions/blog/diversity/2020/why-the-head-of-diversity-is-the-job-of-the-moment.

17 In the wake of COVID-19. See Binyamin Appelbaum, *The Economists' Hour: False Prophets, Free Markets, and the Fracture of Society* (New York: Little, Brown and Company, 2019); Partha Dasguptaa, "The Economics of Biodiversity: The Dasgupta Review," GOV.UK, 2021, https://www.gov.uk/government/publications/final-report-the-economics-of-biodiversity-the-dasgupta-review; Business Roundtable, "Statement on the Purpose of a Corporation," August 19, 2019, https://opportunity.businessroundtable.org/ourcommitment/; and Larry Fink et al., "Sustainability as BlackRock's New Standard for Investing," BlackRock, January 2020, https://www.blackrock.com/corporate/investor-relations/2020-blackrock-client-letter.

18 Ernst & Young surveyed. Ruth Omuh, "A Recent Study Says Some White Men Feel Excluded at Work," CNBC, October 12, 2017, https://www.cnbc.com/2017/10/12/a-recent-study-says-some-white-men-feel-excluded-at-work.html

18 deep-seated consciousness at play. Isabel Wilkerson, *Caste: The Origins of Our Discontents* (New York: Random House, 2020), 125, 180, 183.

Chapter 2: Shedding Messages That Harm Us

25 the first image that comes to mind is not a person of color or a woman. Valerie Purdie-Greenaway and Richard P. Eibach, "Intersectional Invisibility: The Distinctive Advantages and Disadvantages of Multiple Subordinate-Group Identities," *Sex Roles* 59 (September 2008), 371–91, https://doi.org/10.1007/s11199-008-9424-4.

25 Whiteness is part of the "prototype" of leadership. Ashleigh Shelby Rosette, Geoffrey J. Leonardelli, and Katherine W. Phillips, "The White Standard: Racial Bias in Leader Categorization," *Journal of Applied Psychology* 93, no. 4 (2008): 758–77.

29 "The stereotype of the quiet, talented professional." Buck Gee, "A Bamboo Ceiling Keeps Asian-American Executives From Advancing," *New York Times*, October 16, 2015, https://www.nytimes.com/roomfordebate/2015/10/16/the-effects-of-seeing-asian-americans-as-a-model-minority/a-bamboo-ceiling-keeps-asian-american-executives-from-advancing

29 "angry Black woman" at work. Coqual, "Being Black in Corporate America: An Intersectional Exploration," 2019. Melena Ryzik et al., "When Trump Calls a Black Woman 'Angry,' He Feeds This Racist Trope," *New York Times*, August 14, 2020, https://www.nytimes.com/2020/08/14/arts/trump-black-women-stereotypes.html. Daphna Motro, Jonathan Evans, Aleks Ellis, and Lehman Benson III, "Race and Reactions to Women's Expressions of Anger at Work: Examining the 'Angry Black Woman' Stereotype," *Journal of Applied Psychology* (April 2021); summarized in Christina Duran, "Black Women Face Unique Challenges in the Workplace, According to New Study," *Inside Tucson Business*, April 23, 2021, https://www.insidetucsonbusiness.com/business/black-women-face-unique-challenges-in-the-workplace-according-to-new-study/article_48fc7608-9e33-11eb-92a0-db4ff239b6c1.html.

31 women's movements and rights can be limited and controlled. Seema Jayachandran, "The Roots of Gender Inequality in Developing Countries," *Annual Review of Economics* 7 (August 2015): 63–88, https://doi.org/10.1146/annurev-economics-080614-115404; Yoonyoung Cho, Davie Kalomba, Ahmed Mushfiq Mobarak, and Victor Orozco,

"Gender Differences in the Effects of Vocational Training: Constraints on Women and Drop-Out Behavior," The World Bank, 2013, https://documents.worldbank.org/en/publication/documents-reports/documentdetail/882971468272376091/gender-differences-in-the-effects-of-vocational-training-constraints-on-women-and-drop-out-behavior; Kimberly Seals Allers, "Rethinking Work-Life Balance for Women of Color: And How White Women Got It in the First Place," *Slate*, March 5, 2018, https://slate.com/human-interest/2018/03/for-women-of-color-work-life-balance-is-a-different-kind-of-problem.html; Scott Horsley, "'My Family Needs Me': Latinas Drop Out of Workforce at Alarming Rates," NPR, October 27, 2020, https://www.npr.org/2020/10/27/927793195/something-has-to-give-latinas-leaving-workforce-at-faster-rate-than-other-groups

31 Her mother said. Conor Friedersdorf, "Why PepsiCo CEO Indra K. Nooyi Can't Have It All," *The Atlantic*, July 1, 2014, https://www.theatlantic.com/business/archive/2014/07/why-pepsico-ceo-indra-k-nooyi-cant-have-it-all/373750/.

Chapter 3: Carrying Wisdom That Feeds Us

47 There is an Indigenous story. This story is widely repeated; see The Nanticoke Indian Tribe, "The Tale of Two Wolves," 2011, https://www.nanticokeindians.org/page/tale-of-two-wolves.

48 When I watched her TED talk. Kluane Adamek, "The legacy of matriarchs in the Yukon First Nations," TED, November 2020, https://www.ted.com/talks/kluane_adamek_the_legacy_of_matriarchs_in_the_yukon_first_nations.

49 In Clarissa Pinkola Estés's book. Clarissa Pinkola Estés, *Women Who Run with the Wolves: Myths and Stories of the Wild Woman Archetype* (New York: Ballantine Books, 1992)

51 McKinsey published an article. Nicolai Chen Nielsen, Gemma D'Auria, and Sasha Zolley, "Tuning In, Turning Outward: Cultivating Compassionate Leadership In a Crisis," McKinsey & Company, May 1, 2020, https://www.mckinsey.com/business-functions/organization/our-insights/tuning-in-turning-outward-cultivating-compassionate-leadership-in-a-crisis; see also Gemma D'Auria, Aaron De Smet, Chris

Gagnon, Julie Goran, Dana Maor, and Richard Steele, "Reimagining the Post-Pandemic Organization," *McKinsey Quarterly*, May 15 2020, https://www.mckinsey.com/business-functions/organization/our-insights/reimagining-the-post-pandemic-organization.

52 triggered her to move beyond systems thinking. Leyla Acaroglu, "Tools for Systems Thinkers: The 6 Fundamental Concepts of Systems Thinking," Medium.com, September 7, 2017, https://medium.com/disruptive-design/tools-for-systems-thinkers-the-6-fundamental-concepts-of-systems-thinking-379cdac3dc6a.

53 "Anger is an assertion of rights and worth." Soraya Chemaly, *Rage Becomes Her: The Power of Women's Anger* (New York: Simon & Schuster, 2018), 295.

54 Sofiya, a young millennial. I interviewed Sofiya directly, but to learn more about This Same Sky please see Anum Tahir, "This Same Sky on Sustainable Fashion," *Brown Girl Magazine*, August 28, 2020, https://browngirlmagazine.com/2020/08/this-same-sky-on-sustainable-fashion/.

55 "I eat 'no' for breakfast." Courtney Connley, "Kamala Harris on Being Told 'It's Not Your Time' in Her Career: 'I Eat 'No' for Breakfast,'" CNBC.com, November 6, 2020, https://www.cnbc.com/2020/11/02/kamala-harris-on-naysayers-in-her-career-i-eat-no-for-breakfast.html.

Chapter 4: When Your Mind and Body Speak

68 The World Health Organization says. "Social Determinants of Health" (World Health Organization, n.d.), https://www.who.int/health-topics/social-determinants-of-health#tab=tab_1.

69 increased risk for hypertension, infectious illnesses, and a lifetime of physical diseases. David R. Williams, "Race, Socioeconomic Status, and Health: The Added Effects of Racism and Discrimination," *Annals of the New York Academy of Sciences* 896, no. 1, 173–88; and Brea L. Perry, Kathi L. H. Harp, and Carrie B. Oser, "Racial and Gender Discrimination in the Stress Process: Implications for African American Women's Health and Well-Being," *Sociological Perspectives* 56, no. 1 (2013): 25–48, https://journals.sagepub.com/doi/10.1525/sop.2012.56.1.25.

69 "psychological safety." Amy Edmondson, interview with Martha Lagace, "Make Your Employees Feel Psychologically Safe: How to Build a

Fearless Organization," Harvard Business School, November 26, 2018, https://hbswk.hbs.edu/item/make-your-employees-psychologically-safe.

69 "the invisibility syndrome." Anderson J. Franklin, Nancy Boyd-Franklin, and Shalonda Kelly, "Racism and Invisibility: Race-Related Stress, Emotional Abuse and Psychological Trauma for People of Color," *Journal of Emotional Abuse* 6, no. 2 (2006): 9–30, https://www.tandfonline.com/doi/abs/10.1300/J135v06n02_02.

69 medical studies support these findings. Naa Oyo A. Kwate, Heiddis B. Valdimarsdottir, Josephine S. Guevarra, and Dana H. Bovbierg, "Experiences of Racist Events Are Associated with Negative Health Consequences for African American Women," *Journal of the National Medical Association* 95, no. 6 (June 2003): 450–60, https://www.ncbi.nlm.nih.gov/pmc/articles/PMC2594553/; and Rebecca Din-Dzietham, Wendy N. Nembhard, Rakale Collins, and Sharon K. Davis, "Perceived Stress Following Race-Based Discrimination at Work Is Associated With Hypertension in African-Americans. The Metro Atlanta Heart Disease Study, 1999–2001," *Social Science & Medicine* 58, no. 3 (February 2004): 449–61, https://pubmed.ncbi.nlm.nih.gov/14652043/.

75 too much yang, or "doing" energy. When there is too much yin or yang energy, one can become sick. See "What Is Qi? And Other Concepts," University of Minnesota, n.d., https://www.takingcharge.csh.umn.edu/explore-healing-practices/traditional-chinese-medicine/what-qi-and-other-concepts; and Hyun K. Lee, "Hyperthyroidism: Running on Too Much Yang," *Santa Ynez Valley News*, August 4, 2011, https://syvnews.com/lifestyles/health-med-fit/hyperthyroidism-running-on-too-much-yang/article_87a91358-bd64-11e0-83d0-001cc4c002e0.html.

77 peacocks' plumage so beautiful. Peter Fraser, "Pavo Cristatus: The Homeopathic Proving of Peacock Feather," Homœopathic Information Service, 2003, https://www.hominf.org/peacock/peaintr.htm.

77 yum and yuck. Tom Robbins, *Still Life With Woodpecker* (New York: Bantam, 1990), 44.

Chapter 5: The Job Within the Job

84 Pink chair roles are non-revenue-generating responsibilities outside of HR. Ruchika Tulshyan, "Women of Color Get Asked to Do More

'Office Housework.' Here's How They Can Say No," *Harvard Business Review*, April 6, 2018, https://hbr.org/2018/04/women-of-color-get-asked-to-do-more-office-housework-heres-how-they-can-say-no

84 it's the kind of caretaking women naturally assume. Jennifer Kim, "Whose Job Is It to D&I Anyway?: How Should Startups Reward / Compensate Internal D&I Advocates?," Inclusion at Work, October 17, 2018, https://inclusionatwork.co/whose-job-is-it-to-di-anyway/.

85 One of the biggest unpaid burdens. Patricia Faison Hewlin, "Wearing the Cloak: Antecedents and Consequences of Creating Facades of Conformity," *Journal of Applied Psychology* 94, no. 3 (May 2009): 727–41.

87 all this extra work creates a job within the job. For the ways in which Employee Resource Groups can end up as another source of unpaid labor, see Sidney Fussell, "Black Tech Employees Rebel Against 'Diversity Theater,'" *Wired*, March 8, 2021, https://www.wired.com/story/black-tech-employees-rebel-against-diversity-theater/; and Issie Lapowsky, "Leading an Employee Resource Group Is Like a Second Job. Now, One Startup Is Paying for That Work," Protocol, July 24, 2020, https://www.protocol.com/justworks-paying-employee-resource-group.

88 *Cornell Law Review* titled. Devon W. Carbado and Mitu Gulati, "Working Identity," *Cornell Law Review* 85, no. 5 (July 2000), https://scholarship.law.cornell.edu/cgi/viewcontent.cgi?article=2814&context=clr.

91 "heat and light." Dolly Chugh, *The Person You Mean to Be: How Good People Fight Bias* (New York: HarperBusiness, 2018).

94 The sage often feels moved to educate. Adia Harvey Wingfield, "Being Black—But Not Too Black—in the Workplace," *The Atlantic*, October 14, 2015, https://www.theatlantic.com/business/archive/2015/10/being-black-work/409990/.

98 he became a force for social justice. Kurt Streeter, "Kneeling, Fiercely Debated in the N.F.L., Resonates in Protests," *New York Times*, June 5, 2020, https://www.nytimes.com/2020/06/05/sports/football/george-floyd-kaepernick-kneeling-nfl-protests.html.

98 Naomi Osaka and Serena Williams. Elena Bergeron, "How Putting on a Mask Raised Naomi Osaka's Voice," *New York Times*, December 16, 2020, https://www.nytimes.com/2020/12/16/sports/tennis/naomi

-osaka-protests-open.html; "The Black Victims Honoured in Naomi Osaka's US Open Masks," BBC News, September 9, 2020, https://www.bbc.com/news/world-us-canada-54088453; Sonia Elks, "Serena Williams Says 'Underpaid, Undervalued' as Black Woman in Tennis," Reuters, October 6, 2020, https://www.reuters.com/article/usa-race-tennis/refile-serena-williams-says-underpaid-undervalued-as-black-woman-in-tennis-idUKL8N2GX48V; Sean Morrison, "Serena Williams Talks Black Lives Matter, Body Positivity and Female Empowerment," *Evening Standard*, October 6, 2020, https://www.standard.co.uk/news/uk/serena-williams-vogue-blm-body-positivity-and-female-empowerment-a4564146.html; Naomi Osaka, "I Never Would've Imagined Writing This Two Years Ago: Tennis Star Naomi Osaka on Why She Flew to Minneapolis Days after George Floyd's Death—and Why Being 'Not Racist' Isn't Enough," *Esquire*, July 1, 2020, https://www.esquire.com/sports/a33022329/naomi-osaka-op-ed-george-floyd-protests/.

98 LeBron James. Emily Sullivan, "Laura Ingraham Told LeBron James to Shut Up and Dribble; He Went to the Hoop," NPR, February 19, 2018, https://www.npr.org/sections/thetwo-way/2018/02/19/587097707/laura-ingraham-told-lebron-james-to-shutup-and-dribble-he-went-to-the-hoop; LeBron James, Twitter post, June 4, 2020, 2:53 PM, https://twitter.com/KingJames; Uninterrupted, Twitter post, June 4, 2020, 7:19 PM, https://twitter.com/uninterrupted.

99 Viola Davis and America Ferrera are also using their platforms. "WITW L.A. Salon: Viola Davis on Being Told She's 'a Black Meryl Streep,'" Women in the World, 2018, YouTube video, 02:03, https://www.youtube.com/watch?v=Sf0kDGVkVzQ; Nicole Acevedo, "America Ferrera Leads Tijuana Migrant Shelter Visit, Calls for Changes to U.S. Asylum Policy," NBC News, March 11, 2019, https://www.nbcnews.com/news/latino/america-ferrera-leads-tijuana-migrant-shelter-visit-calls-changes-u-n981686.

Chapter 6: Overhauling a Culture of Aggression and Inaction

103 60 percent of WOC feel their companies. Fairygodboss and nFormation, "More Women of Color Are Ready to Leave the

Workforce," March 2021, https://fairygodboss.com/presentation/why-women-of-color-are-leaving-the-workplace.

103 WOC are beginning to take a stand. For example, in June 2020, two Black women who had worked on Pinterest's public policy team alleged that they had experienced discrimination, harassment (including doxxing of their personal information), and gaslighting while at the company. Dr. Timnit Gebru, a Black woman who co-led Google's Ethical AI team before being fired in December 2020, has since spoken out about the unfair treatment faced by WOC in Silicon Valley. And former Bloomberg reporter Nafeesa Syeed sued Bloomberg LP, as well as the white men who managed its editorial committee, for discrimination, alleging a "caste system"; she was joined by another WOC employee a few months later. See Levi Sumagaysay, "'There's a Diversity Grift Right Now': Employees at Center of Racial Controversies at Tech Companies Speak Out," Market Watch, January 2, 2021, https://www.marketwatch.com/story/theres-a-diversity-grift-right-now-employees-at-center-of-racial-controversies-at-tech-companies-speak-out-11609250355; Laura Feinder, "Former Pinterest Employees Allege 'Pure Hell' of Unfair Pay, Racism, and Retaliation," CNBC, June 16, 2020, https://www.cnbc.com/2020/06/16/former-pinterest-employees-allege-unfair-pay-racism-and-retaliation.html; Nitasha Tiku, "Black Women Say Pinterest Created a Den of Discrimination—Despite Its Image As the Nicest Company in Tech," *Washington Post*, July 4, 2020, https://www.washingtonpost.com/technology/2020/07/03/pinterest-race-bias-black-employees/; Kate Duffy, "Pinterest's $22.5 Million Gender Discrimination Settlement Is Another Example of How Black Women Are Ignored, Say Senior Women of Color in the Tech Industry," Insider, December 12, 2020, https://www.businessinsider.com/pinterest-gender-discrimination-women-of-color-tech-racism-black-women-2020-12?IR=T&r=DE; Kari Paul, "Pinterest's $22 Million Settlement With Executive Is a 'Slap in the Face,' Black Former Workers Say," *Guardian*, December 18, 2020, https://www.theguardian.com/technology/2020/dec/18/pinterest-gender-discrimination-lawsuit-black-workers; Zo Schiffler, "Google Fires Prominent AI Ethicist Timnit Gebru," *The Verge*, December 3, 2020, https://www.theverge

.com/2020/12/3/22150355/google-fires-timnit-gebru-facial-recognition-ai-ethicist; and Issie Lapowsky, "For Big Tech Whistleblowers, There's No Such Thing as 'Moving On,'" Protocol, April 15, 2021, https://www.protocol.com/big-tech-whistleblowers. On the Bloomberg lawsuit, see "Bloomberg Fosters Newsroom 'Caste System' with White Men at the Top: Lawsuit," Reuters Legal, August 10, 2020, https://today.westlaw.com/Document/I17ccaec0db5311ea8c4689b27fcbe9be/View/FullText.html?transitionType=SearchItem&contextData=(sc.Default); and "'Current Employee Joins Discrimination Lawsuit Against Bloomberg LP,' WWD," Cohen Milstein, November 14, 2020, https://www.cohenmilstein.com/update/%E2%80%9Ccurrent-employee-joins-discrimination-suit-against-bloomberg-lp%E2%80%9D-wwd.

104 classic microaggression may appear to be a seemingly insignificant slight. Tori DeAngelis, "Unmasking 'Racial Micro Aggressions': Some Racism Is So Subtle That Neither Victim Nor Perpetrator May Entirely Understand What Is Going On—Which May Be Especially Toxic for People of Color," *Monitor on Psychology* 40, no. 2 (February 2009): 42, https://www.apa.org/monitor/2009/02/microaggression; Kristen P. Jones, Chad I. Peddie, Veronica L. Gilrane, Eden B. King, and Alexis L. Gray, "Not So Subtle: A Meta-Analytic Investigation of the Correlates of Subtle and Overt Discrimination," *Journal of Management* 42, no. 6 (October 11, 2013): 1588–1613, https://journals.sagepub.com/doi/10.1177/0149206313506466; and Hahna Yoon, "How to Respond to Microaggressions," *New York Times*, March 3, 2020, https://www.nytimes.com/2020/03/03/smarter-living/how-to-respond-to-microaggressions.html.

104 ruminate more on unpleasant events. Alina Tugend, "Praise Is Fleeting, But Brickbats We Recall," *New York Times*, March 23, 2012, https://www.nytimes.com/2012/03/24/your-money/why-people-remember-negative-events-more-than-positive-ones.html.

104 "emotional tax." "Day-to-Day Experiences of Emotional Tax Among Women and Men of Color in the Workplace," Catalyst, February 15, 2018, https://www.catalyst.org/research/day-to-day-experiences-of-emotional-tax-among-women-and-men-of-color-in-the-workplace/, 4.

108 not coming across as too aggressive or unlikeable. Alisha Haridasani

Gupta, "Likability, Authenticity, Smiles: The Debate Tightrope for Kamala Harris," *New York Times*, October 6, 2020, https://www.nytimes.com/2020/10/06/us/likability-ambition-kamala-harris-debate-mike-pence.html; Joan C. Williams, "How Women Can Escape the Likability Trap," *New York Times*, August 16, 2019, https://www.nytimes.com/2019/08/16/opinion/sunday/gender-bias-work.html.

109 This makes you complicit in acts of violence and discrimination. Ruth Terry, "How to Be an Active Bystander When You See Casual Racism," *New York Times*, October 29, 2020, https://www.nytimes.com/2020/10/29/smarter-living/how-to-be-an-active-bystander-when-you-see-casual-racism.html.

115 more *New York Times* pieces being written. In 2016, the *New York Times* launched its "Race/Related" newsletter; likewise, the *Washington Post* has "Race in America" ("An ongoing series on race in America, examining the movement to end systemic racism and police brutality"). On "Pull Up or Shut Up," Sharon Chuter's campaign to get brands to release the total numbers and seniority levels of Black employees, see Alexa Tietjen, "Sharon Chuter, Ella Gorgla Get Clear About Brand Activism," *WWD*, December 18, 2020, https://wwd.com/beauty-industry-news/beauty-features/sharon-chuter-ella-gorgla-brand-activism-1234672822/; and Gabby Shacknai, "UOMA Beauty's Sharon Chuter Is Holding Brands Accountable with 'Pull Up or Shut Up,'" *Forbes*, June 8, 2020, https://www.forbes.com/sites/gabbyshacknai/2020/06/08/uoma-beautys-sharon-chuter-is-holding-brands-accountable-with-pull-up-or-shut-up/?sh=4703a23070de. In May 2021, the largest statewide employer association in New Jersey established an agreement with the African American Chamber of Commerce of New Jersey to (optionally) attempt to increase Black boards of directors as well as the "development, recruitment and employee advancement" of Black people. "NJBIA and AACCNJ Establish CEO Pledge to Increase Business Diversity," May 4, 2021, https://www.insidernj.com/press-release/njbia-aaccnj-establish-ceo-pledge-increase-business-diversity/.

116 one-third of the US workforce is bound by NDAs. Orly Lobel, "NDAs Are Out of Control. Here's What Needs to Change," *Harvard Business*

Review, January 30, 2018, https://hbr.org/2018/01/ndas-are-out-of-control-heres-what-needs-to-change.

116 cofounder of Lift Our Voices. I interviewed Julie directly, but for more information on Lift Our Voices, see https://www.liftourvoices.org/.

119 numbing ourselves to make our work lives fit into our whole lives? Lakshmi Ramarajan and Erin Reid, "Shattering the Myth of Separate Worlds: Negotiating Nonwork Identities at Work," *Academy of Management Review* 38, no 4 (2013), 621–44.

Chapter 7: The Power of We

129 queen bee syndrome occurs. Cecilia Harvey, "When Queen Bees Attack Women Stop Advancing: Recognising and Addressing Female Bullying in the Workplace," *Development and Learning in Organizations* 32, no. 5 (August 2018), quoted in Jonathan Leake, "Office Queen Bees Bully Female Workers," *Sunday Times*, August 26, 2018, https://www.thetimes.co.uk/article/e8f462b8-a8a1-11e8-8688-ded03633f041.

130 they also don't serve as role models. See Naomi Ellemers, "Women at Work: How Organizational Features Impact Career Development," *Policy Insights from the Behavioral and Brain Sciences* 1, no. 1 (October 1, 2014): 45–54; Naomi Ellemers, Floor Rink, Belle Derks, and Michelle Ryan, "Women in High Places: When and Why Promoting Women Into Top Positions Can Harm Them Individually or as a Group (and How to Prevent This)," *Research in Organizational Behavior* 32 (2012): 163–87; Belle Derks, Colette Van Laar, Naomi Ellemers, and Kim de Groot, "Gender-Bias Primes Elicit Queen-Bee Responses Among Senior Policewomen," *Psychological Science* 22, no 10 (October 2011): 1243–49.

132 white women inherently trust the system, while women of color, especially Black women, do not. Ella L. J. Edmondson Bell and Stella M. Nkomo, *Our Separate Ways: Black and White Women and the Struggle for Professional Identity* (Boston, MA: Harvard Business School Press, 2001), 128–32, 137, 139

136 Girls Who Code. This summary of Girls Who Code is based on my own work with the organization and Reshma Saujani's TED talk. Reshma Saujani, "Teach girls bravery, not perfection," TED, February 2016,

https://ed.ted.com/lessons/teach-girls-bravery-not-perfection-reshma-saujani.

137 Avasara. Soumya Kapoor Mehta and Steven Walker, "Where Are India's Women Leaders?," *Hindustan Times*, March 4, 2021, https://www.hindustantimes.com/opinion/where-are-india-s-women-leaders-101614781421864.html. To learn more about Avasara, see http://www.avasara.in/home.

140 "quietly sustain each other's very survival." Brooke Baldwin, *Huddle: How Women Unlock Their Collective Power* (New York: HarperCollins, 2021), 2.

Chapter 8: How to Play the Game While You Change the Game

141 change from the outside *and* from the inside. Debra Meyerson and Maureen A. Scully call this position "tempered radicalism," and note its salience for WOC in particular. See Debra E. Meyerson and Maureen A. Scully, "Tempered Radicalism and the Politics of Ambivalence and Change," *Organization Science* 6, no. 5 (Sept.–Oct. 1995): 585–600, https://www.jstor.org/stable/2634965.

143 we need a more level playing field. "Women of Color in the United States (Quick Take)" Catalyst, February 1, 2021, https://www.catalyst.org/research/women-of-color-in-the-united-states/.

147 data about pay. Sophie Stuber, "This Company Published Every Employee's Salary Online. Did It Make Pay More Equal?," *Guardian*, February 5, 2020, https://www.theguardian.com/us-news/2020/feb/05/buffer-company-published-every-employee-salary-online-pay-more-equal-gender-gap; Daisuke Wakabayashi, "At Google, Employee-Led Effort Finds Men Are Paid More Than Women," *New York Times*, September 8, 2017, https://www.nytimes.com/2017/09/08/technology/google-salaries-gender-disparity.html; Heidi Lynn Kurter, "How a Former Google Employee's Salary Spreadsheet Inspired This New Movement," *Forbes*, October 20, 2019, https://www.forbes.com/sites/heidilynnekurter/2019/11/20/how-a-former-google-employees-salary-spreadsheet-inspired-this-new-movement/?sh=383084be790a. See also "On the Books, Off the Record: Examining the Effectiveness of Pay Secrecy Laws in the U.S.," Institute for Women's Policy Research,

January 2021, https://iwpr.org/wp-content/uploads/2021/01/Pay-Secrecy-Policy-Brief-v4.pdf.

151 "frozen middle." Jennifer Reynolds terms "the 'frozen middle,' those middle managers who are the gateway between junior to mid-level employees and leaders in the corporation." Jennifer Reynolds, "What Is the Frozen Middle, and Why Should It Keep Leaders Up at Night?," *Globe and Mail*, May 2, 2017, https://www.theglobeandmail.com/report-on-business/careers/leadership-lab/what-is-the-frozen-middle-and-why-should-it-keep-leaders-up-at-night/article34862887/.

151 give more chances to people who look like them. Lauren A. Rivera, "Guess Who Doesn't Fit In at Work," *New York Times*, May 30, 2015, https://www.nytimes.com/2015/05/31/opinion/sunday/guess-who-doesnt-fit-in-at-work.html?_r=0; Allyson Kapin and Craig Newmark, "Tech Aspires to Be a Meritocracy. But It's Only a 'Mirror-Tocracy,'" American Banker, October 18, 2017, https://www.americanbanker.com/opinion/tech-aspires-to-be-a-meritocracy-but-its-only-a-mirror-tocracy. For an explanation of how companies' disinclination to look outside existing, largely white recruiting networks can negatively affect the hiring of diverse candidates, see Quoctrung Bui and Claire Cain Miller, "Why Tech Degrees Are Not Putting More Blacks and Hispanics Into Tech Jobs," *New York Times*, February 25, 2016, https://www.nytimes.com/2016/02/26/upshot/dont-blame-recruiting-pipeline-for-lack-of-diversity-in-tech.html. Google was revealed early in 2021 to have had a tiered ranking system for universities that effectively discriminated against HBCUs; see Nitasha Tiku, "Google's Approach to Historically Black Schools Helps Explain Why There Are Few Black Engineers in Big Tech," *Washington Post*, March 4, 2021, https://www.washingtonpost.com/technology/2021/03/04/google-hbcu-recruiting/. For a summary of the negative impact of "homophily" on diversity in venture capital, see Derek Thompson, "Everybody's in a Bubble, and That's a Problem," *The Atlantic*, January 25, 2017, https://www.theatlantic.com/business/archive/2017/01/america-bubbles/514385/; Kellogg School professor Laura A. Rivera's study of a large professional service organization showed that "employers sought candidates who were not only competent but also culturally similar to themselves in terms of leisure pursuits,

experiences, and self-presentation styles. Concerns about shared culture were highly salient to employers and often outweighed concerns about absolute productivity." Lauren A. Rivera, "Hiring as Cultural Matching: The Case of Elite Professional Service Firms," *American Sociological Review* 77, no. 6 (November 2012): 999–1022, https://journals.sagepub.com/doi/10.1177/0003122412463213.

152 Carla Harris. Carla Harris, "How to Find the Person Who Can Help You Get Ahead at Work," TED, November 2018, https://www.ted.com/talks/carla_harris_how_to_find_the_person_who_can_help_you_get_ahead_at_work.

153 360-degree feedback. Ginka Toegel and Jay A. Conger describe 360-degree feedback as a "typically competence-based survey instrument [that] solicited confidential evaluations from the full range of working relationships a manager possessed—subordinates, peers, and bosses—using a quantitatively based multi-item questionnaire. Each targeted individual in turn received a report summarizing in numerical and descriptive assessments their capability to effectively demonstrate specific competences based on the perceptions of those assessing them." Ginka Toegel and Jay A. Conger, "360-Degree Assessment: Time for Reinvention," *Academy of Management Learning & Education* 2, no. 3 (September 2003): 297–311, https://www.jstor.org/stable/40214201.

157 Saturn car brand. See Peter Valdes-Dapena, "Saturn: Secrets of the 'No-Haggle' Price," CNN Money, September 24, 2006, https://money.cnn.com/2006/09/19/autos/debating_no-haggle/.

160 Ella Baker famously said. Ella Baker, 1977, quoted in Barbara Ransby, *Ella Baker and the Black Freedom Movement: A Radical Democratic Vision* (Chapel Hill: University of North Carolina Press, 2003), 46.

Chapter 9: The Power of Deciding to Stay or to Go

163 changing employers every four years on average. Bureau of Labor Statistics, "Employee Tenure in 2020," September 22, 2020, https://www.bls.gov/news.release/pdf/tenure.pdf.

163 Black women have the highest rate of attrition. Alexis Krivkovich, Kelsey Robinson, Irina Starikova, Rachel Valentino, and Lareina Yee, "Women in the Workplace 2017," McKinsey and LeanIn.Org, October

2017, https://www.mckinsey.com/~/media/McKinsey/Industries/Technology%20Media%20and%20Telecommunications/High%20Tech/Our%20Insights/Women%20in%20the%20Workplace%202017/Women-in-the-Workplace-2017-v2.ashx, 3. This trend has not abated during COVID; McKinsey and LeanIn.Org's 2020 "Women in the Workplace" report notes, "Since the start of Covid-19, Black women are more likely than other employees to think about leaving the workforce because of concerns over their health and safety." This is in addition to the external stresses that may be driving their exits: WOC "are more likely [than white mothers] to be their family's sole breadwinner or to have partners working outside the home during Covid-19. . . . Latina mothers are 1.6 times more likely than white mothers to be responsible for all childcare and housework, and Black mothers are twice as likely to be handling all of this for their families." LeanIn.Org and McKinsey, "Women in the Workplace 2020," September 30, 2020, accessible via https://www.mckinsey.com/featured-insights/diversity-and-inclusion/women-in-the-workplace, 19, 29.

163 two-thirds of over eight hundred WOC surveyed. Fairygodboss and nFormation, "More Women Of Color Are Ready To Leave The Workforce."

163 impacts of COVID-19 and the "shecession." WOC were "disproportionately represented" in industries like accommodation and food services (24.3 percent of whose workers are WOC) and arts, entertainment, and recreation (14.5 percent), both of which experienced mass layoffs due to the pandemic. WOC also disproportionately work low-wage jobs, which are less likely to have paid sick leave or family leave. And large numbers of WOC worked in jobs, some "essential," that may have led some women to quit rather than go to work in a pandemic: ". . . an estimated 60.3 percent of maids and housekeepers, 50.3 percent of nursing assistants, and 45.7 percent of personal care aides are women of color." Jocelyn Frye, "On the Frontlines at Work and at Home: The Disproportionate Economic Effects of the Coronavirus Pandemic on Women of Color," Center for American Progress, April 23, 2020, https://www.americanprogress.org/issues/women/reports/2020/04/23/483846/frontlines-work-home/.

168 For her book. Lisen Stromberg, *Work Pause Thrive: How to Pause for Parenthood Without Killing Your Career* (Dallas: BenBella Books, 2019).

169 Mother Teresa told her. This information is based on my interview with Rani, but to hear more about Rani's story, see "Adobe For All Summit 2019: Rani Mani," AdobeCareers, October 1, 2019, YouTube video, 11:57, https://www.youtube.com/watch?v=5QsxgG_cwMk.

175 Bozoma Saint John, the CMO of Netflix. Bozoma Saint John, interviewed by Maria Menounos, "Strategically Making Your Value Known with Bozoma Saint John," *Better Together*, episode 30, November 11, 2019, YouTube video, 1:32:54, https://www.youtube.com/watch?v=lTfSxauZ0AY.

176 the system on the outside isn't set up for us, either. For more on this subject, see Cheryl Contee, "Advice on Launching a Tech Startup When You're Not a White Man," *Harvard Business Review*, October 11, 2019, https://hbr.org/2019/10/advice-on-launching-a-tech-startup-when-youre-not-a-white-man; and Mary Ann Azevedo, "Untapped Opportunity: Minority Founders Still Being Overlooked," Crunchbase, February 27, 2019, https://news.crunchbase.com/news/untapped-opportunity-minority-founders-still-being-overlooked/.

177 Tiffany's unique ideas and her proprietary platform. Tiffany Dufu, "How to Raise the First $1 Million for Your Startup," The Cru, September 5, 2019, https://www.thecru.com/thoughts/44-how-to-raise-the-first-1-million-for-your-startup.

178 the current venture capital structure rewards quantity over quality. Jennifer Brandel, Mara Zepeda, Astrid Scholz, and Aniyia Williams, "Zebras Fix What Unicorns Break," Medium.com, March 8, 2017, https://medium.com/zebras-unite/zebrasfix-c467e55f9d96.

178 Julia Collins. "Zume Pizza," Crunchbase, accessed March 15, 2021, https://www.crunchbase.com/organization/zume-pizza#section-funding-rounds.

Chapter 10: The New Rules of Power

183 Meghna's little-Brown-girl brain. This information is based on my interview with Meghna, but to hear more about her story, please see Thembi Bheka, "Discover the Rules of Power with Meghna

Majmudar," *She Breaks Thru*, episode 60, podcast audio, https://shebreaksthru.com/episode-60/; for her website, see http://the-permission.com/.

184 Professor Kimberlé Crenshaw. Kimberlé Crenshaw, "Demarginalizing the Intersection of Race and Sex: A Black Feminist Critique of Antidiscrimination Doctrine, Feminist Theory and Antiracist Politics," *University of Chicago Legal Forum* 1989, issue 1, https://chicagounbound.uchicago.edu/cgi/viewcontent.cgi?article=1052&context=uclf. For a recent exploration of the ways in which the term has been used, in and out of the academy, by both the left and right, in the thirty years since Crenshaw coined it, see Jane Coasten, "The Intersectionality Wars," Vox, May 28, 2019, https://www.vox.com/the-highlight/2019/5/20/18542843/intersectionality-conservatism-law-race-gender-discrimination.

184 "it's about changing the very face of power itself." Kimberlé Crenshaw, talk at Women, Power & Peace Conference, September 14, 2007, transcript via Feminist.com, https://www.feminist.com/resources/kimberly_crenshaw.html.

184 As I was researching traditional power, I often saw it talked about in four ways. See Brené Brown, "Dare to Lead," who contrasts "leaders who use power over" with those who use "power with and power to." Brené Brown, "Brené with Joe Biden on Empathy, Unity and Courage," *Unlocking Us*, podcast audio, 2:55, October 21, 2020, https://brenebrown.com/podcast/brene-with-joe-biden-on-empathy-unity-and-courage/.

185 Have you seen MC Escher's. Escher, MC (1938). *Day and Night*.

186 Iyanla Vanzant. Leigh Newman, "Iyanla Vanzant: 7 Things You Don't Know About the Power You've Always Had," Oprah.com, November 9, 2011, https://www.oprah.com/oprahs-lifeclass/iyanla-vanzant-what-you-dont-know-about-the-power-youve-always-had.

188 the Reflected Best Self Exercise (RBSE). To take or learn more about the Reflected Best Self Exercise, see "What Is the RBSE?," Center for Positive Organizations, University of Michigan Ross School of Business, n.d., https://reflectedbestselfexercise.com/about.

189 In their book *Power, for All*. I interviewed Julie Batttilana directly,

but for more on the "Three Pernicious Fallacies" of power, see Julie Battilana and Tiziana Casciaro, *Power, for All: How It Really Works and Why It's Everyone's Business* (New York: Little, Brown, 2021).

191 banding together as a collective. The concept of "power with" rather than "power over" is Mary Parker Follett's. See Kenneth Thompson, Henry C. Metcalf, and Lyndall F. Urwick, eds., *Dynamic Administration: The Collected Papers of Mary Parker Follett* (London: Routledge, 2003 [1940]); and Mary P. Follett, "Community Is a Process," *The Philosophical Review* 28, no. 6 (November 1919), 576–88, https://www.jstor.org/stable/2178307. For a discussion of her context and impact, see Domènec Melé & Josep M. Rosanas, "Power, Freedom and Authority in Management: Mary Parker Follett's 'Power With,'" *Philosophy of Management* 3 (2003): 35–46.

191 States like California. David A. Bell, Dawn Belt, and Jennifer J. Hitchcock, "New Law Requires Diversity on Boards of California-Based Companies," Harvard Law School Forum on Corporate Governance, October 10, 2020, https://corpgov.law.harvard.edu/2020/10/10/new-law-requires-diversity-on-boards-of-california-based-companies/; Julia Boorstin, "Closing the Board Seat Gender Gap," CNBC, March 4, 2021, https://www.cnbc.com/video/2021/03/04/closing-the-board-seat-gender-gap.html.

191 Nicole Anand. I interviewed Nicole directly, but to access her article see Nicole Anand, "'Checkbox Diversity' Must Be Left Behind for DEI Efforts to Succeed," *Stanford Social Innovation Review*, May 21, 2019, https://ssir.org/articles/entry/checkbox_diversity_must_be_left_behind_for_dei_efforts_to_succeed.

192 Ava DuVernay emphasized. "Ava DuVernay's Speech at *Glamour*'s 2019 Women of the Year Awards Must Be Read," *Glamour*, November 11, 2019, https://www.glamour.com/story/ava-duvernay-glamour-women-of-the-year-2019-speech.

193 "Racial capitalism." Cedric J. Robinson, *Black Marxism: The Making of the Black Radical Tradition* (Chapel Hill: University of North Carolina Press, 1983). For a concise discussion of the context of "racial capitalism" within Robinson's work, see Robin D. G. Kelley, "What Did Cedric Robinson Mean by Racial Capitalism?," *Boston Review*,

January 12, 2017, http://bostonreview.net/race/robin-d-g-kelley-what-did-cedric-robinson-mean-racial-capitalism.

194 "On a quiet day, I can hear her breathing." Arundhati Roy, *War Talk* (Cambridge, MA: South End Press, 2003), 74–75.

195 Regional Chief Kluane Adamek's TED talk. Kluane Adamek, "The legacy of matriarchs in the Yukon First Nations."

196 *Thus Spoke Zarathustra: A Book for All and None* by Nietzsche. Friedrich Nietzsche, *Thus Spoke Zarathustra: A Book for All and None*, trans. Walter Kaufmann (New York: Modern Library, 1995).

197 "Courage is the most important of all the virtues." Maya Angelou, interview, n.d., YouTube video, 01:49, https://www.youtube.com/watch?v=xWTXkVPhEXU.

INDEX

ABOUT THE AUTHOR

DEEPA PURUSHOTHAMAN is a cofounder of nFormation, a company for women of color by women of color. nFormation provides brave, safe, new space for professional women of color. Deepa is also a Women and Public Policy Program Leader in Practice at the Harvard Kennedy School.

Prior to this, Deepa spent more than twenty years at Deloitte and was one of the youngest people and the first Indian American woman to make partner in the company's history. Deepa was also the US managing partner of WIN (Women's Initiative), Deloitte's renowned program to recruit, retain, and advance women.

Deepa has degrees from Wellesley College, Harvard Kennedy School, and the London School of Economics. She speaks extensively on women and leadership. She has been featured at national conferences and in publications including *Bloomberg Businessweek*, HuffPost, and *Harvard Business Review*. She is also an Aspen fellow.

Deepa and her husband, Manoj, live in Los Angeles with their dogs and an endless list of home renovations. If you want to know more about Deepa's current projects please visit her website at deepapuru.com.